LOVE
SCENTS

LOVE SCENTS

How Your Natural Pheromones Influence Your Relationships, Your Moods, and Who You Love

BY

MICHELLE KODIS

WITH DR. DAVID MORAN

AND DEBORAH HOUY

A DUTTON BOOK

DUTTON
Published by the Penguin Group
Penguin Putnam Inc., 375 Hudson Street, New York, New York 10014, U.S.A.
Penguin Books Ltd, 27 Wrights Lane, London W8 5TZ, England
Penguin Books Australia Ltd, Ringwood, Victoria, Australia
Penguin Books Canada Ltd, 10 Alcorn Avenue, Toronto, Ontario, Canada M4V 3B2
Penguin Books (N.Z.) Ltd, 182–190 Wairau Road, Auckland 10, New Zealand

Penguin Books Ltd, Registered Offices: Harmondsworth, Middlesex, England

First published by Dutton, an imprint of Dutton NAL, a member of Penguin Putnam Inc.

First Printing, October, 1998

10 9 8 7 6 5 4 3 2 1

LIBRARY OF CONGRESS CATALOGUING-IN-PUBLICATION DATA

Kodis, Michelle.
Love scents : how your natural pheromones influence your relationships, your moods, and who you love / Michelle Kodis, with David T. Moran, and Deborah Houy.
p. cm.
Includes bibliographical references and index.
ISBN 0-525-94333-1
1. Pheromones—Popular works. 2. Chemical senses—Popular works. 3. Emotions—Popular works. I. Moran, David T. II. Houy, Deborah. III. Title.
QP572.P47K63 1998
612.8'6—dc21 98-18126
CIP

Printed in the United States of America
Set in Garamond Light
Designed by Eve L. Kirch

This book is printed on acid-free paper. ♾

Dedicated by Michelle Kodis to:
Robert, Joan, and Steven Kodis, my beloved family

Dedicated by Dr. David T. Moran to:
The human sixth sense, which has been guiding our species along the way to becoming the wonderful creatures we are all the time, even though we've just "discovered" it

Dedicated by Deborah Houy to:
Nick Houy

CONTENTS

FOREWORD

In the pages of this fascinating and timely book, you will learn about how humans communicate through chemical signals called pheromones, and how these silent but remarkably effective communication tools provide each of us with a vast range of information about the people we meet in casual passing, the people who become our friends, and the people who become our intimate partners.

As a scientist who has spent many years studying human pheromones, I can say with confidence that this field of research holds great promise for a number of notable, and potentially revolutionary, scientific breakthroughs and developments. Indeed, in a relatively short amount of time in the scientific continuum, pheromones have evolved from being known and understood as chemical signals that function only in insects and animals into what are now known to be the crucial, and perhaps most basic, components of human interaction. In the 1960s, when I was an anatomy professor at the University of Utah and engaged in studies of human skin, to have suggested that pheromones belonged in the human repertoire of interaction would have been unheard of, maybe even ridiculed. Today, however, thanks to detailed and innovative scientific studies, pheromones have finally been awarded their deserved place in the language of human relationships.

In the past few years, pheromones have made their debut in the world media, earning headlines in magazines and newspapers and receiving wide coverage on television and radio. To date, there has not been a popular book that focuses on the newest and most cutting-edge pheromone studies and theories that stand poised to illuminate more of the mysteries of these invisible but powerful chemical messengers. *Love Scents* homes in on this latest research. The book presents its information in an easy-to-read, personable style for the layperson. Learning more about pheromones helps all of us to better understand each other and our own actions, motivations, and responses. In the future, pheromone technology may be present in our lives in the form of a new class of pharmaceuticals, a topic which is presented in *Love Scents*. This is a book to be read with delight—and with hope.

David L. Berliner, M.D.
Pherin Pharmaceuticals, Inc.
Menlo Park, California
February 1998

ACKNOWLEDGMENTS

Michelle Kodis:

Along with Deborah Houy, I want to thank Lisa Ross of The Spieler Agency for guiding this book through its numerous stages with a demeanor that was not only delightful, but heartening, as well. Our appreciation goes to editor Deirdre Mullane, whose remarkable range of talents helped to shape an idea into a book. I wish also to recognize Dr. Susanna Hoffman, Susan Viebrock, and Dr. Susannah Smith for their friendship and encouragement, and Dr. Howard Donner for his inspirational words. Dr. David Berliner, a true gentleman, answered our questions patiently and took our calls promptly, and Drs. Louis Monti-Bloch, Clive Jennings-White, and Margaret Johns were very generous with their time and knowledge. Thanks also to Jean Koch, Virginia Erickson, Susan Smilanic, Elizabeth Plamondon, and Dianne Dumanoski. Lastly, my love and appreciation to Jim Pettegrew and Sasha for the nourishing sunset and full-moon walks through the aspen trees of our mutual paradise.

Dr. David T. Moran:

I would like to acknowledge the help, encouragement, and fellowship—both academic and personal—of my colleagues Drs. David Berliner, Louis Monti-Bloch, Clive Jennings-White, Carter Rowley, Bruce Jafek, and Larry Stensaas, whose research on human pheromones has produced a body of knowledge that will greatly benefit the human race. Furthermore, I would like to thank Ms. Michelle Kodis for doing the lion's share of the work on this book; for mastering a complex body of emerging knowledge in a remarkably short time; and for communicating the fascinating subject of pheromone biology in such a clear and interesting manner.

Introduction

It is the body that melts with love, freezes with fear, trembles in anger, and reaches for warmth and contact.

—Alexander Lowen, *The Betrayal of the Body*

This book is about a monumental scientific discovery: Humans possess a true sixth sense that works in concert with the other five senses of smell, taste, hearing, touch, and sight. Before this revelation, our lives were governed by a knowledge of the well-documented five senses; indeed, it seemed as if those sensory gifts were more than enough to keep us in touch with our surroundings, feelings, desires, fears, the depths of our emotional cores.

In recent years, however, scientists have been homing in on a previously undiscovered sensory system that performs its work quietly and is responsible for our most basic human reactions and interactions. This system is triggered by the mysterious chemical messengers we now know as pheromones. According to Dr. Louis Monti-Bloch, a scientist at the University of Utah whose studies into how humans perceive pheromones have received international acclaim, "There is overwhelming evidence that humans have a sixth sense."

A sixth sense? The phrase itself is not new to many of us. For hundreds of years, people have shared stories of gut feelings and intuition, of a deeply felt "sixth sense" that spoke to them. How many times have you referred to this mystical entity in casual conversation? Perhaps you've said something like, "I don't know why I avoided that situation. My sixth sense must have been telling me something."

This sense is active. It works busily to influence our lives in ways we never before would have thought possible. As you will discover in your journey through *Love Scents*, the sixth sense silently asserts its opinions in your decisions to love, hate, marry, take a job, bond with your newborn baby, befriend a neighbor, or avoid a coworker in the corporate hallway.

You live according to the directional road maps provided by the intimate connection of your six senses, but your sixth sense is wired directly to the most primitive region of your brain, an area that sits apart from the well-developed "thinking" brain and is responsible for your emotions and basic bodily functions. This circuitry guides you through your day, sending signals and opening and closing gates in response to each situation you face. When you feel receptive and warm toward someone, the gates let in positive signals that convince you to explore a new relationship; when you feel wary of another, they shut.

The revolutionary discovery of the physiological pathway for the sixth sense can be compared to recalling suddenly something you've known for many years, a detail that has registered deep within your brain and has been present all the time, just waiting patiently to be called forth.

What You Already Know

Love Scents will tell you what you already know innately. Why, then, would you spend precious time reading about something that is supposedly familiar? It's simple: We believe you've never looked at your sixth sense in the ways we will suggest. As we've said already, you probably refer to this sense in terms of intuition or gut feelings or an otherworldly perception that guides you through life.

The real sixth sense, however, is quite different from what you might think it to be, and that is why we have delved so deeply into this subject. We believe that once you have read *Love Scents*, you will begin to view your life—your interactions with other people, your gut feelings about random events and situations, your love life, the romantic fling you had years ago or yesterday—from a new vantage point.

The sixth sense lives deep within your primitive brain and it affects you in amazing ways. It also affects the people around you—your spouse, your children, your lover, your coworkers, the bank teller, the teenager who bags your groceries.

As you learn about this sense, you will discover a deeper knowledge of what it means to be human. This awareness will unlock some of the mysteries of everyday life and shed light on your motivations and on those of the people around you. It is information that will allow all of us to tap into a more profound sense of who we are.

The Sixth Sense at Work

The first thing that strikes you about Susan and Richard is that they are so physically different. She is classically beautiful, poised, blond. Richard probably wouldn't turn your head. He is slightly pudgy, somewhat on the short side, and balding. His profile is anything but movie-star chiseled. But Susan and Richard's attraction is as powerful as a Herculean magnet. Susan says she fell in love with Richard the first time they met.

He introduced himself, I introduced myself, and I felt as though the floor had fallen away from my feet, Susan recalls. *As they say in the love songs, my knees went weak.*

Richard nods in agreement. Because Susan is so beautiful, Richard had doubted that she would find him attractive. Before their formal introduction, Richard had fantasized from afar about a relationship with Susan, but he didn't think a woman like her would express even a shred of interest in a man like him. But the energy between them is electric. Both of them have had a number of long-term, even passionate, relationships, but this one is clearly different.

It's unlike anything I've experienced in my life, Richard says. *It's pure chemistry. Pure magic.*

When Jane and Susannah, both psychologists, met at a conference, the two women felt an instant camaraderie. They chatted about life, love, work, and children as longtime friends often do. They shared similar upbringings, educational backgrounds, and senses of humor. But what surprised them the most was the ease they felt together.

There was none of the uncomfortable stuff you sometimes get when meeting a person for the first time, Jane remembers. *It was as if I'd met someone with whom I felt completely connected.*

Years later, they are still close friends. Their bond has continued to deepen, although they now live in different parts of the country. What keeps them close? Susannah thinks it is the memory of the initial "hitting it off."

Unlike the friendship that developed effortlessly between Jane and Susannah, Nicholas's first meeting with Ian was tainted with an immediate feeling of discomfort that was confusing at the time. Nicholas, a physician, found it impossible to connect with his colleague.

Here I am, a logical, thinking human being, and I'm making assumptions about a person just because I felt uncomfortable the first time we met, Nicholas recollects. *Still, even when I tried to be logical, to remember that first impressions don't always stick, I couldn't shake the feeling that this man and I were not going to be close colleagues.*

Nicholas's hunch was on the mark. His first encounter with Ian had ended abruptly. From that day on, their interactions were strained and eventually the men began to avoid each other. They even went so far as to sit at opposite ends of the conference table during staff meetings.

It's simple, says Nicholas, *our chemistry was off.*

Think about it. Have you ever been bowled over by an attraction or instant dislike to someone that seemed to come out of nowhere?

One man told us a certain woman's scent left him feeling "intoxicated." *My brain just shuts off when I'm around her,* he said. *Her scent is that powerful to me.*

Is it her perfume, her lotion, her laundry soap? we asked. He claims this "smell" has nothing to do with man-made fragrances or perfumes. *It's something deeper,* he explained.

During the course of researching *Love Scents,* we talked to many people who are in relationships we will call "chemically in tune." Based on interviews with people of all ages, backgrounds, and situations, we came to the conclusion that attraction follows similar patterns.

It usually begins like this: We experience an instant, urgent desire to be close to a person. Getting closer increases the electricity, and a simple touch can send a current of desire through both people. The "smell" of the other person is often described as "sweet." One woman, a writer in her forties, said she seeks out a specific patch of skin on her husband's face to sniff. Doing so, she says, practically puts her into a trance. When she and her husband, a commercial airline pilot, have a disagreement or she feels a need to connect with him, she will move in close and inhale his scent. *It works every time,* she says.

A little background on this story: When this woman, whom we will call Elizabeth, met her husband, it seemed they had little in common. He grew up on a working cattle ranch in eastern Washington, she in Manhattan and its environs. She knew about art history, enjoyed the fast life of New York City, shopped in upscale boutiques, and had risen to a top position at one of the world's largest banks. He knew about riding horses, playing guitar and composing music, flying 747s across the oceans, and sailing—and he didn't care much about couture or city life. By all appearances, they seemed destined to hit the brick wall of incompatibility. But that was not to be the case.

Elizabeth and her husband have been married nearly a decade, and their partnership is the envy of many of their friends. Now living in rural Colorado, they are still passionately in love and their successful relationship is a model for what can happen when two people are chemically suited to each other. Despite disparate backgrounds and a former divergence, rather than a convergence, of lifestyle, this couple works, and they have the spark to prove it.

Our Methods

The researching and writing of *Love Scents* proved both interesting and challenging. In addition to spending many hours interviewing scientists who work in the up-and-coming arena of pheromone research, we conducted numerous "field interviews." In other words, we spent a lot of time just talking to people and asking them about their past and present relationships. Most of our subjects were delighted to divulge the romantic details of their relationships and told their stories freely.

As we sat down and analyzed these conversations, we kept noticing constants that mark the most passionate and sometimes most successful pairings. On the flip side, we also talked to people about their negative romantic and work experiences. People react strongly to other people, that we know. What we found interesting is that strong reactions appear on both sides of the coin—whether you are falling madly in love or running away from someone who makes you uneasy. Regardless of the scenario, the "constant" we speak of is *chemistry*.

The stories presented in *Love Scents* are real, although the names of the people in question have been changed to honor requests of anonymity. We believe that the most interesting stories are those based on real-life circumstances and events; however, some stories are composites, drawn from several accounts to add color and to emphasize a point.

The saga of pheromones is a still-unfolding one. We begin by giving you an overview of pheromones and then go on to examine how animals use chemical communication. From there, we move into a more scientific mode and guide you through the anatomy of pheromone perception. We'll take a brief but fascinating tour through the human brain and discuss theories of how pheromone perception fits into the complex workings of the other five senses. Next, we lighten things up a bit and delve into the irresistible topics of sex, love, and lust, followed by a chapter that sheds light on even more wonders of pheromones. We'll discuss pheromone-laden products like perfume and then look at the future of pheromones as the leading scientists and we see it.

During the course of our research, we unearthed a wealth of information that we have used to ground the reader in the basics of pheromone research and development. As we sat down and outlined *Love Scents*, we realized it would benefit from a combination of true-life stories and hard science, and we have intertwined those threads whenever possible. Fortunately, our scientific collaborator, Dr. David T. Moran, is one of the leading pheromone experts in the world. His involvement in this book helped us greatly and opened doors to information that might have been difficult or impossible to obtain had our way not been paved. We spent several years researching pheromones and collecting material related to the study of human interaction. We went to libraries, universities, medical schools, bookstores, and computerized databases to find everything we could that either focused on pheromones or was related to the topic. Our searches proved fruitful and provided us with a breadth of material that gives this book a significant and broad scope of coverage.

Have You Experienced the Chemistry of Pheromones?

Before we begin, the following questionnaire will help you think about the chemical conversations in your life. If you are currently not involved in an intimate relationship, recall a past encounter or relationship when answering the following questions. Keep in mind that the word *relationship* does not always imply a sexual liaison. In this context, it could also refer to a friendship or an acquaintance.

1. When I met my partner/friend, I was aware of an instant, undefined chemistry flowing between us:

- Agree Very Strongly
- Agree Somewhat
- Disagree

2. I have had encounters in which I became aware immediately that my response to the person in front of me was "off" for some reason:

- Agree Very Strongly
- Agree Somewhat
- Disagree

3. My partner/spouse and I have a strong sexual attraction, and the chemistry seemed to exist from the moment we met:

- Agree Very Strongly
- Agree Somewhat
- Disagree

4. I have had experiences where I felt strongly that my intuition, gut feelings, or what I have referred to as my "sixth sense" came into play:

- Agree Very Strongly
- Agree Somewhat
- Disagree

5. I have ignored a gut feeling or intuition about someone only to discover later that my instincts were correct:

- Agree Very Strongly
- Agree Somewhat
- Disagree

6. I am often aroused and excited when in the presence of my sexual partner:

- Agree Very Strongly
- Agree Somewhat
- Disagree

7. In general, I would avoid a relationship with someone if I felt our chemistry wasn't in sync:

- Agree Very Strongly
- Agree Somewhat
- Disagree

8. I have ended or would end an intimate relationship that lacked chemistry:

- Agree Very Strongly
- Agree Somewhat
- Disagree

9. I have declined or would decline job offers and other career advancements and opportunities based on the strength of my gut feelings and intuition:

- Agree Very Strongly
- Agree Somewhat
- Disagree

10. For some unexplained reason, I feel sexier and more receptive during the warmer months of the year:

- Agree Very Strongly
- Agree Somewhat
- Disagree

If your answers are mostly "Agree Very Strongly":

You are tuned into the powers of your sixth sense and the silent communication of pheromones. You are cognizant of the impact of pheromones on your everyday life, and you consider the role of chemistry seriously in all of your relationships.

If your answers are mostly "Agree Somewhat":

You are aware that chemistry is present in your life and you pay attention to it, but you could become more tuned into how pheromones and your sixth sense engage you in chemical conversations with other people. How can you do this? By consciously evaluating events and situations in your life, which will then help you to hone your sixth sense and discover the true power of pheromones.

If your answers are mostly "Disagree":

You have much to learn about pheromonal communication. To become better able to decode the chemical conversations in your life, you might want to keep a diary of your encounters with other people, documenting your first impressions and how they made you feel. After a while, you'll begin to notice the subtle moments that occur in your interactions with your fellow humans.

Now, read on.

1

Chemical Conversations

The meeting of two personalities is like the contact of two chemical substances: If there is any reaction, both are transformed.

—C. G. Jung

A happily married man meets a woman at a party and within minutes is imagining her in his bed. He is shocked by his immediate rush of feelings. *What's happening?* he wonders. *She's not as beautiful as my wife and yet I can't keep my eyes—or my thoughts—off her.*

An employee at a large corporation feels uncomfortable in the presence of a certain division manager and makes every effort to avoid her peer. Her intensely visceral reaction is a mystery. *This person hasn't done anything to me, and we don't even really know each other,* the employee tells herself. *So what could be causing this response?*

What do these two scenarios have in common? They describe what can happen in the presence of *pheromones*, chemical signals that tell us important things about the people we encounter every day. Pheromones can ignite an instant and perhaps even startling attraction, quietly advise us to avoid a threatening or overly aggressive person, or bond us to our best friends.

Invisible but Powerful

Pheromones are odorless molecules that are produced in the body and enter the world by wafting off the skin. They also float up from the recesses of the sweat glands and linger in strands of hair. Each unleashed pheromone molecule is packed with information about your sexual desires, your level of aggression, the attributes of your immune system, and more. Every pheromone carries your one-of-a-kind chemical "signature," which is as unique as the swirls of your fingerprint.

Pheromones tell you about your neighbor, your best friend, your coworker, the man who reads your electric meter, the person who sits next to you on the bus. Your pheromones are odorless and invisible, and as they glide through the air, they carry with them vital information about you. Although you might not be aware of it, pheromone messenger molecules whirl off your body and into the air twenty-four hours a day. Their target is other humans. And, when a pheromone hits its target, it delivers its message.

For example, consider the powerful impact of first impressions. A friend introduces you to his cousin and you register an instant dislike. At this point, your logical brain will most likely speak up and suggest in its rational way that you shouldn't form such a strongly negative opinion in mere seconds. Nevertheless, you can't erase the nagging thought: *I don't like this person*. Most people have engaged in this delicate battle between logic and the more intangible effects of emotion and intuition.

This is but one of the many mysteries of pheromones. Pheromones bypass the logical, thinking brain and affect the center of the primitive, emotional brain. Many of us live in societies that place merit on controlling the emotions and other primitive responses, but pheromones' effects are always present to remind us that such control is not always possible.

There is a fascinating story in the psychological literature about a woman who was visiting Oxford, England, when she found her world turned on end. She was standing by a river when a group of young male students walked by. As the boys passed, the woman had the unmistakable and instant feeling that one of these youths was the infant son she had given up for adoption only a few hours after his birth. The young man said later that he had experienced a similar visceral, gut reaction to the female "stranger" standing by the water. As he walked by her, a thought flew into his brain: *That's my mother!*

What makes this story particularly powerful is that the woman and the young man did not look like each other, so the recognition was not based on a similarity of physical traits. How did these people from vastly different worlds know that one of their own was in the vicinity?

While hugged in the cozy, wet warmth of the womb, the fetus uses pheromones to communicate with its mother in a quiet code of chemical signals. This chemical link survives the baby's birth so that a mother can identify her infant not only by his smell but also by his pheromones. Pheromones also help guide the infant's mouth toward the nourishment of its mother's breast. Could this chemical bond be so strong that a mother separated from her baby soon after his birth would be able to identify him many years later?

In the vast landscape of human emotion and physiology, scientists continue to be amazed by what people can do; thus, recognizing a son years after giving him up for adoption is not all that far-fetched. After all, humans are chemical creatures. We may believe we've evolved beyond the need to communicate with subconscious chemical signals, but the fact is we do just that all the time.

Many of us shower or bathe daily, shampoo our hair, launder our clothes, and scent our bodies with perfumes and colognes. Because we control how we present ourselves to the world, we might believe that we have mastered our subconscious messages. In general, we live according to the credo of our large, developed brains, the brains that have given us the gift of higher thought processes. We are *human*. To hint that we are driven by chemical impulses is a discomfiting notion.

But, scientists have discovered that our best intentions to control what we broadcast into the world may be futile. Could it be that we have much in common with animals and insects, whose cues for sexual reproduction, dominance, and aggression spring directly from pheromonal communication?

What provokes gut feelings about certain situations and people? Why are we attracted to one person and repelled by another? What announces our sexual availability, our aggressive attitudes, our plays for dominance on the sports field or in the corporate boardroom? What makes us love James instead of Robert, Christine instead of Patricia, Sam instead of Martin? In *A Natural History of the Senses*, writer and naturalist Diane Ackerman calls pheromones "the pack animals of desire." It now seems possible that these "pack animals" are largely responsible for hooking us up with people whose chemistry incites desire or some other positive emotion in us and for keeping us away from people whose chemical signals we deem unappealing.

In his book *Society and Solitude*, Ralph Waldo Emerson said, "Society exists by chemical affinity, and not otherwise." Emerson was not referring to pheromones, but to the centuries-old human conflict between solitude and spending time with other people. Still, this quotation can be used to illustrate our point: A kind of built-in chemical affinity, which could be related directly to pheromones, defines human interaction. That may be why people who live in isolation often have weaker-than-average immune systems and diminished psychological health. When you choose one lover over another, avoid an acquaintance, pursue a romance even though it might not be the wisest thing to do, or nurture the bond between yourself and your newborn baby, new research suggests that you are subconsciously following the ancient rules of pheromones; in other words, you are being influenced by your sixth sense.

Nineteenth-century German philosopher Friedrich Nietzsche summed it up best when he said, "All of my genius resides in my nostrils." That genius, it seems, is innate in all humans.

Early Suspicions

While the topic of human pheromones is relatively new to the world of science, for centuries people have suspected that *something* was occurring between themselves and other people that couldn't be explained easily and fully with the sensory definitions of the day. Both popular and academic literature is peppered with anecdotes of human interactions that appear to involve pheromonal communication.

William C. Agosta, a leading animal pheromone researcher at New York City's Rockefeller University, recounts such an anecdote in his scientific investigation *Chemical Communication: The Language of Pheromones*. The story hints at the inexplicable powers of subconscious pheromonal signals that are still being uncovered layer by layer. According to reports published in the psychosexual literature of the nineteenth century, a young Austrian peasant was very skilled at attracting the ladies. Before entering his favorite dance hall, the peasant would tuck a handkerchief into the cave of his armpit; once securely in place, the cloth would become soaked—perfumed, you could say—with sweat. When the peasant's female dancing partners became overheated with activity, he would remove the handkerchief from his armpit and use it to wipe their faces. It was a thoughtful gesture, but a larger question looms: Did the peasant enjoy steady success at dances because he unknowingly (but quite smartly) delivered potent blasts of his pheromones to the local girls? Armpits—not usually what comes to mind in a discussion of romance and sex—are indeed sexy because they are one place on the body where pheromones are produced in abundance.

Perhaps modern-day society will embrace a practice that reached the height of its popularity during Shakespeare's time. It was common then for a woman on the prowl to tuck a bit of peeled apple under her arm. When the fruit had absorbed her perspiration, the lady would offer her specially prepared "love apple" to the suitor of her choice, who would sniff the pheromone-imbued tidbit in an odd expression of foreplay. If the suitor liked the scent (and the pheromones) of the apple, intimacy was sure to follow.

Nineteenth-century French poet Charles Baudelaire posited that the soul resides in the erotic sweat. He had never heard of pheromones, but was he referring unknowingly to human chemical interaction and attraction? And what of the comment of French novelist Joris-Karl Huysmans, who claimed the odor emanating from a woman's underarms "easily uncaged the animal in man." Is this unleashing of raw passion really the work of invisible, odorless pheromones?

Of course, we can't forget to mention the now-famous correspondence between Napoleon and his empress, Josephine. In one of his letters, Napoleon informs his beloved that he will be returning home in a few days. He signs off with this succinct request: *Please, do not bathe*. Napoleon understood that his lover's powerful sexual perfume (perfume laced with pheromones, we now know) would have been lost with the interference of soap and water.

Louis XIV, the French Sun King, stank horribly because he shunned baths, calling them unhealthy. Louis also chased smelly women. The smellier the woman, the more lustful Louis's response. While the king's pungent bouquet may have offended the sensitive noses of his court, he was in fact listening to and obeying his true human nature: By not bathing, Louis kept his pheromones concentrated on his skin. His attraction to unbathed women was probably stoked by the fact that they, too, didn't wash off their sweat—and their pheromones—every day.

Journalist Bill Moyers once asked mythology scholar Joseph Campbell the question, "Why do you think we fall in love with one person and not another?" Campbell's response directs us once again to the possibility of pheromone-induced love: "It's a very mysterious thing, that electric thing that happens, and then the agony that can follow. The troubadours celebrate the agony of love, the sickness the doctors cannot cure, the wounds that can be healed only by the weapon that delivered the wound."

The "agony of love"? The "sickness the doctors cannot cure"? Can the simultaneously exhausting and exhilarating experience of being lovesick, of being impaled by the arrow of Eros, be reduced to a chemical reaction in the brain?

A New Science Begins

The story of pheromones began to unfold in the 1950s, when scientists Peter Karlson and Martin Lüscher created the word *pheromone* to refer to the invisible chemical communicators used by the lower orders. (A combination of the Greek words *pherein* ["I carry"] and *hormon* ["to excite"], *pheromone* means "I carry excitement.") But even a new word added to the lexicon didn't spark immediate studies into how pheromones might affect humans; it was thought that pheromones served a purpose only in the animal and insect worlds, and research money was distributed accordingly. Since the discovery of pheromones in moths nearly four decades ago, chemical communication in insects and mammals has led to a number of significant revelations, which we will explain in chapter 2. In short, animal pheromones incite behaviors ranging from "Let's mate!" to "Watch out—I'm defending my territory and will become aggressive if you ignore my warning."

Pheromones are chemicals that one individual emits to elicit responses in another individual of the same species. A pheromone provides species-specific chemical communication and elicits a neurophysiological response that results in an alteration of sensual behavior. Eventually, as the knowledge of the role of pheromones in the behaviors of animals and insects began to take shape, some scientists wondered whether humans might also communicate with pheromones.

Are pheromones the bottom-line creators of love and lust, the orchestrators of chemical magic between two people? As they try to shed light on that question, the players in this exciting field of scientific research are modern-day detectives attempting to unravel the mysteries of human interaction, in a new twist on the quest for the true meaning of amour. Are we aware of our chemical conversations? Not consciously. But the physical and emotional effects of these messages (pheromones shrieking "I'm angry!" or "I want you!") ricochet through our lives. These unconscious conversations may influence us when we spurn a suitor, begin a friendship, reject a job offer, or fall passionately in love with a soul mate. You could say this book is a detective story, one that is mixed with the tempting ingredients of human love and attraction.

A knowledge of pheromones can give you a better understanding of your own behavior. For example, you meet someone you like instantly. The prospect that pheromones might be involved could take your recounting of the meeting from the typical "We have similar hobbies and value systems and therefore are suited to each other" to the more instinctively human "I like his pheromones!"

How do pheromones function in the company of smell, taste, hearing, sight, and touch? Human interaction involves myriad sensory cues, but scientists now think pheromone reception may be the granddaddy of the senses, the chemical communications that travel in unbridled celerity to the unconscious control centers of the brain, where they elicit a number of responses.

We should note here that pheromones can be supplanted by the force of other sensory information. For instance, you might form an impression of someone based on his or her physical appearance; you would be registering an image based on sight. Perhaps you are pleased by the sound of a person's voice; you would then base your impression on sound as you took the stranger's vocal resonations into the delicate curves of your inner ear's cochlea. But if that same person moved closer to you, your impressions would be influenced heavily by pheromones.

When the *New York Times* asked the late astronomer Carl Sagan, "So what can't science explain?" he responded, "Well, I am deeply in love with my wife. I do not claim that I can understand my being in love with her on grounds of pheromones, which certainly explain sexual attraction in moths. Seems to me there's more to it in this human relationship. But that doesn't mean that it will never be explained."

Even if the explanations are not fully formed at this time, the current thinking among a growing number of scientists is that pheromones might very well represent the essence of human communication. As pheromones function in the company of our other senses, they provide the information necessary to discern good from bad, desirable from undesirable, aggressive from passive. They can announce the undeniable tingle of gut feelings and the wake-up call of intuition. They can tell us when to advance, retreat, or just sit tight. Jonas Salk, inventor of the polio vaccine, said, "The intuitive mind tells the thinking mind where to look next." Thanks to innovative research into human pheromones, it appears that the true sixth sense is becoming more clearly understood.

Pheromone Basics

It has been known for some time that animals and insects have no control over the chemical signals that tell them what to do. New research suggests that humans, too, process the cues delivered by pheromonal messages, and that these messages travel straight to our primitive brains.

Pheromones are thought to originate from steroids produced in the body; when those steroids flow to the skin, they are metabolized and transformed into pheromones. The most potent pheromone centers are located in the groin, the armpits, and, interestingly, in the narrow strip of skin between the base of the nostrils and the upper lip—the nasal sulcus. This may be why people kiss—to more effectively sniff out each other's pheromones. Pheromones are processed by two tiny organs inside the base of each nostril. Kissing, then, is an ingenious way to get the nose closer to a pheromone-rich region of the body.

The human skin is also the springboard for some pheromones. One inch of skin contains an average of 625 sweat glands, 90 sebaceous (oil) glands, approximately 65 hair follicles, nearly 20,000 sensory cells, and 24 feet of blood vessels. The largest organ in the human body, the skin contains three distinct glands that are thought to generate pheromones: the apocrine, sebaceous, and eccrine glands.

The main function of the eccrine glands is to produce an odorless watery perspiration and move it to the surface of the skin. When this odorless liquid evaporates from the skin, we are kept cool and protected from overheating. The apocrines produce a strong scent and vary greatly in size and occurrence among individuals. They are dotted around the body but are concentrated in the armpits, the groin, and the nipple area. The apocrines produce the instantly recognizable smell of body odor, which gets its strong scent from the mingling of apocrine gland secretions with bacteria living on the skin. The apocrines kick into high gear during puberty, which explains why adolescents experience a sudden onset of powerful body odor. The sebaceous glands are located throughout the body as well, but are concentrated on the face, forehead, ears, and scalp. These glands produce an oily substance that provides an ideal breeding ground for the bacteria living on human skin.

After these glands (with the apocrines leading the way) produce pheromones, the molecules then make their way to the surface of the skin. In just one hour, humans can shed as many as one thousand skin cells per square centimeter of skin; this rapid sloughing of skin protects the body against bacteria. When the skin sheds, bacteria have a harder time penetrating through the protective barrier and entering the body.

Skin flakes, each composed of many thousands of skin cells, are packed with pheromone molecules. When a skin flake detaches from the skin and starts traveling through the air, the pheromone molecules can then be released to do their work. This is a very basic explanation of a complicated physiological process, but it serves to illustrate in a simple manner how pheromones move from your body to the pheromone-receiving systems of the people with whom you come into contact.

Before pheromones reach the skin, however, they must be

manufactured in the body. One theory of pheromone production involves dehydroepiandrosterone (DHEA), which sex therapist and researcher Dr. Theresa Crenshaw calls "the mother of all hormones."

DHEA and Pheromones: A Possible Link

DHEA is a sex hormone with an impressive repertoire. As a possible precursor to human pheromones, it is a vital component of everyone's body chemistry.

In addition to its role in pheromone production, DHEA performs a number of important functions within the human body. For example, DHEA is thought to boost the immune system, improve brain function, act as an antidepressant, reduce levels of cholesterol in the blood, and encourage bone growth. This wonder-hormone also has the ability to boost a person's sex drive, or libido. During orgasm, DHEA levels in the brain soar.

Classified as a steroid hormone, DHEA metabolizes into pheromones mainly through action in the adrenal glands, which are located on the kidneys. DHEA is also produced in the testicles, the ovaries, and the brain.

In *The Alchemy of Love and Lust*, Dr. Theresa Crenshaw says DHEA is "a most versatile hormone that first and foremost has the potential to manipulate our sexual selections through smell." However, now that the sixth sense has been discovered, we can say that our sexual selections are driven not only by smell, but also by pheromone reception. DHEA levels peak when a person is in his or her midtwenties—the time in the life cycle when mate selection and sexual activity are heightened as the two sexes prepare to have children. Levels begin to drop when adults reach their midthirties.

While we might not be consciously aware of the effects of DHEA, each of us was exposed to the hormone in the womb. A growing fetus is enveloped in DHEA. In fact, the fetus produces DHEA in concentrations that eclipse other hormones, including the prominent sex hormones estrogen and testosterone.

The VNO: The Pheromone Sensor

One of the most exciting topics now under investigation is how humans process pheromonal messages. Scientists have been eager to determine whether humans possess the anatomical conduit for transmitting these messages to the brain. Since pheromones exist, the human body must contain a pathway that moves these chemical messages to the brain. A component of that conduit has been found in a small organ located in the nose—the vomeronasal organ, or VNO.

Let's go back in time for a moment. The year is 1703 and a Dutch military surgeon identified in the literature as "Ruysch" is peering into what remains of the smashed and bloodied face of a wounded soldier. As Ruysch attempts to piece the soldier's broken face into a recognizable whole, his eyes register a tiny slit inside the man's nose, in the front portion of the nasal septum.

Further investigation by Ruysch reveals an identical pit located symmetrically on the other side of the nasal wall. The soldier's nose contains two delicately fleshy pits, but for what purpose? Fascinated by something he'd never before seen despite years of performing facial surgeries, Ruysch looked more closely. He documented his finding.

Ruysch's description of the nasal feature caused little excitement and even Ruysch himself did not investigate it further. A century later, in 1811, Danish scientist Ludwig Levin Jacobson attached his own name to the pit, which he had seen in animals but not in humans. He called it Jacobson's organ, a name that has endured to this day in the world of animal science.

In the early twentieth century scientists had, for the most part, concluded that the human vomeronasal organ existed in only a very small percentage of people and, when present, did not function. Convinced this was the case, writers of medical textbooks usually dismissed the human VNO as vestigial, an evolutionary relic.

A vestigial structure or organ, such as the human appendix, is one that occurs or persists as a rudimentary or degenerate structure. This label seemed to seal the VNO's fate until just a few years ago, when experiments revealed that the VNO does indeed have a life of its own. Two scientists at the University of Colorado Medical School in Denver had some new and innovative ideas regarding the VNO. Dr. Bruce Jafek, an ear, nose, and throat surgeon, and Dr. David Moran, a cell biologist and electron microscopist, thought the VNO might exist and function in most, if not all, humans. Their research, as well as the work of other scientists, will be described in detail in chapter 3.

The vomeronasal organ is anatomically complex but elegant. In mammals, the structure is also referred to as the end organ of the accessory olfactory system. The human VNO opening ranges in size from 0.1 to 2 millimeters. While it often requires magnification to be seen, it is occasionally visible to the naked eye.

Imagine peering into a nose, using a flashlight to deliver a pinpoint of light. If you were to angle the light toward the area in the front one-third of the nasal cavity, you might be able to see a small, fleshy pit, or what appears to be an opening. Investigation with a microscope would reveal an opening that leads to a tube lined with columnar cells. These cells are classified as pseudostratified columnar epithelium, and what's intriguing about them is that they are not found anywhere else in the human body—they are unique to the VNO. Research has also focused on the nerve pathway that links the VNO to the hypothalamus, a primitive structure of the brain that sits near the base of the skull.

The hypothalamus, often called the "brain's brain," is a fascinating structure. Neurology textbooks call the hypothalamus the seat of emotional expression, and this is an apt description because it regulates everything from fear to aggression to sex drive. The hypothalamus also controls the body's water and salt balances, blood pressure, sugar and fat metabolism, endocrine (hormone) function, appetite, and temperature, and is instrumental in a number of other psycho-physical states, such as anxiety. Without a properly functioning hypothalamus, humans would be emotionally bankrupt. And scientists are discovering that the hypothalamus plays a critical role in the mysterious dance of pheromones by receiving chemical messages through the vomeronasal organ and then triggering the appropriate emotion or response.

Dr. David Berliner, president of the pheromone research company Pherin Pharmaceuticals, a biotechnologist and a former anatomy professor whose research on human pheromones is a key source for the information in this book, has taken the link between the VNO and the hypothalamus even further by testing ways to deliver drug therapies in the form of synthetic substances acting on the VNO. In general, pheromone communication works like this: Messages taken into the VNO travel to the hypothalamus, reaching the brain in fractions of a second. Once at the emotional core of the brain, the message carried by the pheromone elicits the designated response. Thus, like animals responding to pheromonal cues as part of daily existence, we, too, react to pheromones without thinking.

From Obscurity to the Spotlight

Five years ago, few people were talking about human pheromones. Regardless of what was happening in a few labs around the country, people certainly weren't discussing pheromones over lunch, and the popular press hadn't yet realized that pheromones make for interesting feature stories and enticing headlines.

Now that the existence of human pheromones has been confirmed and their role in our behavior has been documented, it seems clear that these chemical communicators have important things to tell all of us.

What does it all mean? The possibilities are enormous. Scientists are now investigating how pheromones and their derivatives may help you lose weight, get a better night's sleep, slip into a mood conducive to romance, heighten your libido, combat the effects of anxiety and panic disorders, and treat premenstrual syndrome and breast and prostate cancers.

Whether pheromones enter our lives—and our vomeronasal organs—through our loved one's sweet-smelling sweat, a newly purchased bottle of perfume, or drugs and other therapies that could change medicine, we can be certain they will continue to affect us on many levels.

The concepts introduced in this chapter of *Love Scents* have prepared you for your upcoming journey into the fascinating world of pheromones. As you read on, you will encounter a new way of viewing and assessing human behavior, and begin to appreciate the intricate connection between the five senses and the sixth sense.

2

Pig Breath and Other Animal Wonders

Be a good animal, true to your instincts.

—D. H. Lawrence

Long before scientists suspected the role of pheromones in human interaction, they were aware of their place in animal courtship and behavior. As you are about to see, the pheromonal capabilities of some animals are quite wondrous and, at times, surprising.

Getting love is remarkably easy for a male pig. He exhales a waft of breath on which floats a complex symphony of pheromones telling the world (or the nearest sow), "I'm ready to mate—who's available?" If she's in heat, the sow will respond without what have become the preludes to human romance. Flowers, chocolates, a candlelit dinner with wine are not necessary. The sow accepts amour without courtship.

When the boar's breath, which via the saliva contains the pheromone 5-alpha-androstenol, reaches the female, she assumes a mating posture, giving her body over to the pure and singular message that it's time to reproduce her kind. She arches her back and slips into lordosis, lifting her genitals toward the male. If she doesn't get a hug or a cuddle afterward, no worry. Upon completing the sex act, the sow and the boar (in all of our references, a boar is an uncastrated male pig) have accomplished their goal: propagation of the species.

Pheromones also explain the strange attraction of pigs to truffles. Researchers at the Technical University of Munich and the Lübeck School of Medicine were curious about why a sow will root passionately for hours in search of a single truffle. They unearthed an amazing fact: Truffles send out a chemical come-hither that is identical to 5-alpha-androstenol. Truffles produce even greater quantities of the chemical than do pigs. No wonder the poor, confused female truffle pig cannot help but follow her nose to the source. As Carl Sagan and Ann Druyan wrote in *Shadows of Forgotten Ancestors*, "Since truffles are fungi, in which steroids play key sexual roles, perhaps tormenting sows is just an accidental side-effect—or perhaps it serves the function of inciting pigs to dig so the spores are spread more widely and the Earth is covered with truffles."

Hog farmers have capitalized on the fact that the sow is highly receptive to the pheromones in the male pig's saliva and breath. "Boar Mate" is used by farmers to coax female pigs into lordosis without putting an actual male in front of them. This makes artificial insemination more effective and profitable: one spray of the stuff and the sow is smitten.

Pigs provide but one example of how nonhuman animals use pheromones to communicate. The topic of animal pheromones has been researched extensively; we address it briefly in these pages, but the notes at the end of the book contain sources for readers who wish to learn more.

The Role of Pheromones in the Animal World

Humans—particularly females—expect courtship or a show of romance as a precursor to sexual intimacy, but the lives of animals are often governed by decidedly unromantic pheromones.

Whether the goal is to mate, control the female's reproductive cycles, affect gestation and onset of puberty, aggressively defend a nest, lay a territorial boundary, send a warning signal, establish a relationship of dominance/submissiveness, identify members of one's family, or act in a maternal manner toward offspring, pheromones are the key components of the chemical communications that drive animal behavior.

How and where these pheromones are produced and released varies greatly from creature to creature. For example, the single-celled amoeba *Dictyostelium discoideum* produces a pheromone, acrasin, that attracts others of its kind. Hamsters and other rodents have pheromones in their vaginal secretions. Female hamsters produce and release the strong pheromone *aphrodisin*, which stirs up sexual interest and behavior in the male. Dogs, horses, deer, camels, and a number of other grazing ungulates also have concentrated levels of pheromones in their urine. So powerful are canine pheromones that a bitch in heat can lure males from miles away with her pheromones. Pigs, we have just learned, load up their saliva with pheromones.

Some animals produce pheromones in specialized glands located in the genital and anal areas of their bodies and in the sebaceous glands of their skin. Throughout history, the odiferous substances secreted by these glands in the musk deer, civet cat, beaver, and muskrat have been sought after and used in perfumes (the evolution of perfume is discussed in chapter 7). In some species of insects, specialized "message" glands exist solely to produce pheromones. For example, the male cockroach has a gland on its abdomen that secretes a pheromone designed to incite mating behavior in the opposite sex.

Like humans, other mammals and reptiles possess a vomeronasal organ that processes pheromone signals. The location of the VNO varies among species, but most often it is near the oral or nasal cavity. (Most insects carry their pheromone receptors on the delicate branches of their antennae.) Mammals, amphibians, and reptiles all possess olfactory and accessory olfactory systems composed of separate neural pathways that regulate a number of chemosensory functions. These pathways branch out to separate functioning systems. One, which detects odors, begins in the olfactory region in the nasal cavity and sends information via the olfactory nerve to the olfactory bulb in the brain. The other pathway begins in the vomeronasal organ and terminates in the accessory olfactory bulb in the brain. While the main olfactory system is responsible for such basic behaviors as grooming and feeding, the accessory olfactory system, or the vomeronasal system, receives and processes pheromones.

It's important to note here that pheromones are *species-specific*; that is, a pheromone from a silkworm moth will only attract another silkworm moth and not a butterfly.

Discovering Animal Pheromones

The Johns Effect

In 1978 Margaret Johns, then a doctoral candidate at Rutgers University in the school's Institute of Animal Behavior, designed and conducted an experiment to explore how environmental influences can affect ovulation in rats. In many cases, female animals ovulate after copulation, but Johns found that the female rats in her study ovulated when exposed, even briefly, to the urine of a male rat; copulation was not required. The finding was exciting, but Johns had to answer a greater question: Why? She suspected it might be connected to the sense of smell, but her instinct told her another mechanism could be involved. She went back to her laboratory and discovered that when the female rats were separated from the males, even if by a short distance, the females did not ovulate.

Meanwhile, graduate student Anne Mayer had developed a way to seal off the animal's VNO, thereby preventing it from detecting pheromones. Johns already knew about the VNO, but she had not tailored her experiments around that organ. Johns decided to use Mayer's VNO cauterization technique on her rats. She closed the female rats' VNOs and brought in the males. Even close contact with the males and their urine failed to induce ovulation in the females whose VNOs had been closed. This discovery showed a link between the animal VNO and the mammalian reproductive system. It was, Johns told us, "a thrilling discovery." Her finding is now referred to as the Johns Effect.

Butenandt's Bombykol

The role of pheromones in the insect world had been investigated nearly two decades before Margaret Johns discovered the importance of the rat VNO in the late 1970s. In 1959, German organic chemist Adolph Butenandt identified the female silkworm moth's sex pheromone, the airborne substance that attracts male moths to her at mating time. He called the pheromone *bombykol*, a label derived from the moth's Latin name, *Bombyx mori*.

Butenandt accomplished his task by first gathering more than half a million female moths. From there he began the delicate process of extracting the pheromone from the moths' abdomens, where it is produced in specialized glands. All told, the moths produced less than 7 milligrams of bombykol, but that was enough for the study of the pheromone's molecular structure.

The female silkworm moth can release ten-billionths of a gram of bombykol every second. When a male moth is exposed to the pheromone, he beats his wings, a sign that he is excited. So potent is bombykol that male moths, which can detect as few as one or two molecules of it from significant distances, can follow the female's pheromone trail for miles. Such long-distance communication also occurs among other insects whose pheromones are considered to be some of the most biologically active compounds ever documented. Extreme volatility of pheromone molecules (meaning they are easily evaporated) translates into detection at incredibly low levels. In this category, however, the moth is unbeatable at sensing minute wafts of pheromones.

Looking for Love

Pheromones can improve the reproductive success rates of a number of animals. One study performed by scientists at Damascus University examined how a pheromone extracted from the wool of rams could improve the reproductive rates of adult merino mutton ewes. The researchers found that females treated with ram pheromones experienced enhanced ovulation. They also conceived more readily and showed significantly improved birth rates, and their lambs were healthier.

The examples that follow are designed to show you the scope of "love" pheromones in the animal world. While there are many more examples to relate, each of these serves to illustrate how pheromones work in courting and mating rituals.

On the Fly

Researchers have found that the male fruit fly's semen contains pheromones that can do a number of things to affect and even control the female's reproductive cycles, from inducing her to ovulate so that his sperm, and not another fly's, will fertilize her egg, to quelling her urge to copulate so that she will have less desire to mate with another male. These "anti-female" pheromones improve the odds that the male's sperm will make it to the female's egg, thus ensuring that his progeny will live to carry on the family line.

Certain genetic strains of female fruit flies exude an intoxicating pheromone that makes the spindly-legged males weak at the knees. The chemical makeup of the female's essence, which resembles oil of citronella (another powerful magnet for the male), attracts him instantly.

Chemical Ticks

The tick uses several different pheromones as it goes about its daily business. Without pheromones, the tick (many of which have no eyes) would live in a dark world. One pheromone—appropriately called an assembly pheromone—helps the tick attract the company of other ticks. When the tick is lonely, it releases this pheromone onto the surfaces it touches. Other ticks who happen to pass over these chemically paved surfaces stop what they are doing and cluster together. By promoting strength in numbers, the assembly pheromone is thought to protect the tick.

Mating among ticks is facilitated by the sex pheromone 2,6-dichlorophenol (2,6-DCP). The female releases 2,6-DCP as she feeds on a host. In response, males feeding nearby become excited. They stop eating, detach themselves from the host and go in search of the female, who, filled with blood, can now produce eggs.

Beetle Juice

The female Douglas fir beetle emits pheromonal signals designed to attract males. First, she locates the fir tree, her preferred home, by its scent and bores a hole into it. Once ensconced, she sends out a pheromone that announces her sexual availability. When male beetles detect the pheromone, they begin to fly toward its origin. Upon arriving at the female, the males shut down her pheromone production with a special acoustic signal that essentially pulls the plug on her sexual radio. Then the males start to emit a chemical that jams the pheromone receptors of any other beetles that may be contemplating a move to this particular tree.

Cockroach Attachments

Perhaps the greatest skill of the cockroach is its ability to reproduce in large numbers. In his classic *Life on a Little-Known*

Planet, Howard Ensign Evans wrote that the sex lives of cockroaches reveal "the secret of their success in keeping the world populated with their kind."

Indeed, cockroaches are structurally built to be reproductive wunderkinder. While it is sometimes necessary for the male cockroach to have physical contact with the female before interest in mating is sparked, there are instances in which the female sends out a pheromone that attracts the male even if he's not in her immediate vicinity.

For example, in the American cockroach, the female launches her pheromone—periplanon-b—to tell the male to prepare for sexual intercourse. He approaches, facing her head-on. The two then begin to touch their antennae together in a method Evans calls "fencing."

Following a brief period of antennal foreplay, the male turns away from his love interest. He then raises his wings to release his own sex chemicals that drift from the glands on his back. If the female is still inclined to mate, she will climb on the male's back and begin to eat his chemical exudate. These chemicals induce her to assume a mating posture. When she does so, the male unmasks his genital organs, which he uses to attach himself to her. Says Evans, "She is 'hooked' very literally for the hour or two required for copulation."

In the animal world, the male who succeeds in his sexual advances often exhibits the most "masculine" traits. In the Tanzanian forest cockroach, this can occur to a chemical extreme—but it doesn't guarantee necessarily that the guy will get the girl. University of Kentucky entomologists found that some male cockroaches of this species secrete high concentrations of a pheromone component that puts them in charge and sends the submissive males, whose quantity of the pheromone chemical is not as potent, far down the ladder of superiority. When in the company of submissive males, the dominant cockroaches will kick, bite, and run into them in attempts to knock them over.

The scientists studied the molecular structure of the pheromone in question and found that it is composed of three different chemicals: two that lead to aggressive behavior and one that is responsible for submissive behavior, or "groveling," as the entomologists describe it. When the "dominant" pheromones were placed on the heads of the submissive males, they, too, became aggressive and bullying. But, the question to ask is this: How might the male cockroach benefit at all from a "groveling" pheromone? One theory is that the females are put off by male bullies, even though they display the usually desirable masculine traits, and that the groveling pheromone helps to temper that behavior, which the females perceive as dangerous to themselves and to their kin.

Hamster Dances

The male golden hamster will not mate if he cannot detect the pheromones from the female of his species. A series of chemical transactions must occur before mating begins. First, the female in heat will emit a pheromone announcing she is ready to mate. In response, the male greets the female by sniffing the scent glands on her head. She begins to warm up into a posture that broadcasts her intent to submit.

Following the female's "yes" signal, the male will start to lick her flank; in doing so, he homes in on another scent gland located on the side of her body. From there, he moves to her rear, still licking and sniffing. Only once he detects the special pheromone that directs him to mount her does he do so.

Quill Pigs

The porcupine (its name means "quill pig"), with its nearly thirty thousand quills, has perhaps the greatest physical disadvantage in the courting ritual. But the porcupine, one of North America's largest rodents, makes good use of its pheromones by making urine an integral part of its love game.

In *A Natural History of Sex*, Adrian Forsyth recounts an observation of the porcupine's dating behavior: "The male usually coaxes the female to the ground, where he will rear on hind legs and tail while emitting low vocal 'grunts.' He then proceeds to spray the female with bursts of urine from a rapidly erecting penis, and after wrestling chases, vocalization and more urine showers, coitus is effected."

Snowshoe Ballet

Adrian Forsyth also details the great pains the male snowshoe hare takes to lure the female to his bed. No slouch in the entertainment category, he performs powerful jumps into the air in a valiant attempt to convince the female to mate with him. The ritual doesn't end with ballet-style leaps and bounds, however. While in the air, the male sends a stream of urine onto the female.

Instead of being repulsed by the male hare's behavior, the female perks up. Upon sniffing his urine, she becomes informed of his sexual prowess because the substance reveals the potency of his pheromones and his levels of the sex hormone testosterone.

Lemur Light Sticks

Male ring-tailed lemurs use pheromones to attract the attention of females dozing in trees. The males coat their long tails with their own personal pheromone, and then proceed to wave their tails at each other in scenes reminiscent of the light-stick fight between Luke Skywalker and Darth Vader in *Star Wars*. These ostentatious gesticulations have a purpose: The guy with the most "attractive" tail gets the girl.

Snakes

Alfred Hitchcock's lushly dark movies are frightening because of the unexpected twists played out under seemingly normal circumstances. This method applies just as well to the ploys of the red-sided garter snake (*Thamnophis sirtalis parietalis*), undoubtedly the Hitchcock of the reptilian mating scene. Let's say you're a male snake vying for female attention amid a sea of other males focused on the same goal. Perhaps there are only a few females, making the need to stand out in the crowd even more critical.

Some male red-sided garters are equipped with an ingenious way of winning this reproductive game: They turn themselves into pseudofemales by dressing up in a chemical mask designed to mimic the opposite sex. Adrian Forsyth calls them "chemical transvestites."

During mating, a female snake may become buried under a slithering mass of lusty males. The smartest male will be able to emit a "female" pheromone that causes the other males to become confused and start searching for the source of this "feminine" sex attractant; they may even attempt to mate with "her." While the other males are moving about in befuddlement, the "female" male moves in on his target—the true female.

Acting Boarishly

When competing for the attention of a female, the male wild boar will spar with other males. He will lay his scent on trees and other objects to mark his territory, incorporating urine and dung into the staking of his claim. After this has been accomplished, the male will advance toward the female. If she doesn't turn and flee, he will begin to rub his snout over her body, urinating and grunting at the same time. As the boar prances around the female, he delivers to her an invisible bouquet of pheromones. Many times, the tactics work and the female allows the male to mount her.

Horsing Around

When a stallion gets a whiff of a mare in heat, he slips into what appears to be overly aggressive behavior: He harasses the female, kicking and biting at her until she urinates. When this happens, the stallion knows he is on the way toward accomplishing his goal—copulation.

He moves closer to the mare, preparing to snort her urine by curling his lips back and flaring his nostrils in a gesture called *flehmen*. By performing flehmen, the stallion enhances his sensitivity to the mare's taste and moves her pheromones into his VNO. A "reading" of this sort tells him the state of the mare's reproductive health and the propriety of his advances.

Camels, deer, zebras, giraffes, and rhinos also use the urine-sniffing method and lip curl to determine the sexual availability of their females. The rhinoceros is the most skilled at urine (and pheromone) spraying. A rhino can send a twelve-foot stream of urine to broadcast his presence. This action is akin to fencing off an area and putting up a sign that says KEEP OUT: TRESPASSERS WILL BE PROSECUTED.

A Bull (Elephant) in a China Shop

It's next to impossible to not notice a bull elephant in a state of sexual excitement. Indeed, the male is so aroused and prone to fits of extreme aggression that some zoos won't keep male elephants on the premises. There is a one-word reason for this prejudice: *musth*.

Musth means "intoxicated" in Urdu. A bull elephant experiencing musth, which can last up to two months, does act as if he's consumed too much liquor and is pressing for a bar brawl with anyone who will participate.

The cause of musth is the bull's raging testosterone levels, which are pumped up to fifty times normal. With rivers of sex hormone coursing through his veins, the bull elephant is unpredictable and edgy—and definitely in the mood for some attention from a female (cow) elephant.

What the male elephant really wants is to find a cow in heat; when he does, he will watch over her like a harem master, waving off with ear-splitting bellows and threatening gestures any lower-ranking males who might attempt to mate with her. The cow seems to enjoy the male's displays; to her, he is the most powerful and therefore the one with whom she will mate.

A male in musth exhibits several unmistakable signs, including a preoccupation with marking anything in his path with a gooey secretion produced in the temporal glands on the sides of his face. He will graffiti the local vegetation with this secretion, either by rubbing the glands against trees or shrubs or by touching his trunk to the secretions and then smearing the substance around. He will also spend a great deal of time autographing his territory with his pungent, pheromone-laden urine.

Like stallions, male elephants display flehmen activity when exposed to the urine of the female. Instead of flaring his lips, though, the elephant coats the end of his trunk with the female's urine and then tastes the liquid. This taste test facilitates copulation between the male and the female, as it keys him in to her reproductive cycle.

A remarkable study conducted by scientists at the Oregon Graduate Institute of Science and Technology in Portland revealed that some female insects and elephants release the same pheromone to alert males of their reproductive readiness. This elephant pheromone, (Z)-7-dodecen-1-yl acetate, is the same chemical also used by many insects, "from cabbage loopers to dingy cutworms." Elephants and insects do not read each other's mating messages because the elephant produces the pheromone in much larger quantities. Also, both creatures' mating pheromones are combined with other chemicals that mask the basic underlying molecular structure.

Survival by Chemistry

Think about how useful it would be for humans to be able to send out a warning or alarm pheromone, one that sufficiently tells the person on the receiving end exactly what needs to be said. Some animals have clever ways to ward off predators or announce when trouble is in the neighborhood.

The sea anemone *Anthopleura elegantissima* uses a warning pheromone called anthopleurine. When the anemone is injured, it goes into convulsions and releases a jet of anthopleurine into the water to announce to other members of its "family" that danger is nearby. The other anemones respond by flexing their tentacles, retracting them into their mouths and then clamping their mouths shut. This chain reaction, which allows the anemones to garrison their fragile anatomical structures before the invader arrives, occurs in a matter of seconds.

The anemone can also protect itself from one of its primary predators, the sea slug *Aeolidia papillosa*. The slug feeds on the succulent flesh of the anemone's tentacles, but it doesn't know that its meal is most likely tainted with anthopleurine. Thus, upon finishing its dinner, the slug is full of the anemone's potent warning pheromone, which it is not equipped to metabolize. Without the ability to metabolize anthopleurine, the sea slug is a moving vessel filled with anemone alarm pheromone. When the pheromone-filled slug approaches other anemones, it announces its arrival by unwittingly releasing the ingested pheromone into the water. The anemones are warned that they, too, are at risk of being eaten.

The earthworm *Lumbricus terrestris* produces a warning pheromone when it is in danger or injured. It then secretes a substance that infuses its body with a taste that is unpleasant to many of the animals that feed on it.

Many invertebrates use pheromones to help them journey along the treacherous path of survival. Without the ability to use this chemical communication, their futures would be uncertain.

Once spring feeding begins in the waters of the Arctic, the female Arctic barnacle (*Balanus balanoides*) emits a pheromone that tells her eggs food is plentiful and that it is therefore safe for the larvae to hatch. This hatching pheromone, triggered when the female begins to eat, keeps her eggs safely intact until she can be certain her offspring won't starve. Some crabs also use hatching pheromones to control the activity of their eggs.

Marking Territory

Pheromones can be used to announce an animal's presence. Dogs are perhaps the most familiar of the markers; they travel from tree to tree and bush to bush, urinating on anything that has the potential to hold their individual pheromone codes, which are then read by other dogs passing by. While many animals use pheromones to mark territory and determine who belongs in the backyard, two examples illustrate this point nicely.

Watch a domesticated feline etch itself into its environment. With serpentine grace, the animal moves through the room, rubbing the sides of its cheeks on furniture, on the floor, on the legs of any people who happen to stand or sit in the animal's path.

As the cat glides and sways to the rhythm of its own internal tune, it is depositing pheromones secreted by glands located on both sides of its face. The cat marks a territory, leaving behind its unique imprint and posting an unmistakable sign to any new cats who may appear on the scene: THIS IS MY HOUSE—AND DON'T YOU FORGET IT!

Beavers also communicate with a complex directory of pheromone and odor signals. Individual animals emit pheromones to identify themselves to others in the area. Despite the complexity of these chemical signals, beavers are so adept at translating them that they can accurately single out an individual that is not a member of their family. In short, beavers can put their noses on strangers, passersby, and trespassers who do not belong.

Female Dominance

Some female animals can use pheromones to create dominant-submissive relationships with other females of their species. A dominant female can even go so far as to suppress ovulation in her inferior sisters so that she is the only one to give birth! A female rodent wishing to assert her status within the pack will emit a pheromone designed to tell the other (often younger) females that she is the dominant one and that they are to assume a submissive demeanor.

Biologist David Abbott of the Wisconsin Regional Primate Research Center in Madison has documented dominant-submissive behavior patterns among groups of female marmosets. In marmoset communities, only one female gives birth, often to twins. The other females help her to care for her babies, finding food and baby-sitting when needed. Abbott found that the dominant female—the one who gives birth—uses a pheromone that causes the "inferior" monkeys' ovaries to shrink and stop releasing eggs.

Although pheromones are the main cause of ovulation suppression in marmosets, Abbott found that visual cues and a good deal of pushing and shoving (which amount to bullying) are also a factor. Abbott housed submissive females together; each of these monkeys ovulated about once every ten days. When these submissive females were exposed to the pheromones and sideways looks of a dominant female, they took three times as long to ovulate. Other animals that use ovulation-suppressing pheromones include the jackal, mongoose, and naked mole rat.

Copycats

To ensure their success on earth, some plants and animals engage in clever ploys with pheromones. Copycat pheromones are marvelous examples of the "first me, then you" attitudes often displayed in this world.

Perhaps one of the most amazing examples of copycat pheromones is found in tropical orchids. Orchids need wasps to spread their pollen. Consequently, some Mediterranean and South American varieties produce flowers similar in appearance to a female wasp or bee. To encourage the spreading of its pollen, the orchid sends up a plume of fabricated wasp sex pheromones, essentially turning itself into a female wasp. Confused male wasps attempt to mate with the flowers. As if intoxicated by the heady aroma of the female of their kind, the males fly from one orchid to the next, stopping for a little "love" at each flower and thereby spreading the pollen.

The tiny bolas spider has a creative way of making sure its main food source, the moth, is always in plentiful supply: It mimics the pheromones of the female moth. When male moths sense the pheromone, they mistakenly think the producer is the female of their species, not a predator. They fly toward the source—the spider—and upon arriving find themselves trapped in a deadly web. The bolas spider is a skilled pheromonal cross-dresser. It can mimic the pheromones of sixteen different moth species!

Some beetles, arachnids, and millipedes can emit a pheromone identical to that of ant larvae. This chemical disguise allows them to live inside ant nests, where they dine inconspicuously on the ants' precious eggs and larvae.

Pheromone Experts

The honeybee and the ant provide two of the most complete examples of pheromone communication, and they are among the most sophisticated users of pheromones known to the scientific world.

The Queen Bee

The honeybee (*Apis mellifera*), whose enthusiasm for its craft is matched only by the enthusiasm humans have for eating the amber-colored confection it produces, follows the hard-and-fast rules of pheromones as it labors in the hive.

In the world of the honeybee there is room for only one queen. And grand she is. Presiding over a kingdom that can hold as many as sixty thousand subjects, the queen uses her pheromones to control the drones (males) and the workers (infertile females). As the only fertile female of the lot, the queen has quite a job description, for she must keep the hive full of bees. She mates and lays her eggs in the cells of the hive. A prolific reproducer, she can lay a staggering fifteen hundred eggs per day.

The queen's absolute power is made possible by her special "primer" pheromones, which only she produces. Primer pheromones are more complex than ordinary pheromones that serve to attract a mate, and their effects have a broader scope. The queen uses them to force her drones to mate with her and to prevent them from going off and finding a new queen. Her pheromones tell the rest of the hive that business is proceeding as usual or when danger is in the air. Primer pheromones motivate the lower-ranking bees to take care of the queen's offspring, and they inhibit the sexual maturation of the female worker bees so that the queen is the only one who can reproduce. The queen's pheromones are so powerful that the workers will slack off in their absence. When the worker bees are no longer able to detect the queen's pheromones (as often happens when she gets old), they will assume it is time to prepare for a new ruler and will start constructing special "queen cells" in which the up-and-coming royal highness will be prepped for her duties.

A deceased honeybee broadcasts its misfortune by emitting a "death" pheromone called oleic acid that instructs the other bees in the hive to remove its carcass. So powerful is oleic acid that a live bee smudged with the pheromone will find itself an unwilling participant in its own funeral and subsequent removal from the hive. Even the untouchable queen, if doused with the pheromone, would face the same indignity.

Pheromones also serve to protect the bees' living space. When a beehive is threatened by an intruder, the bees will collectively release a pheromone that incites all of them to attack and sting the unwelcome visitor.

Wasps use pheromones to single out members of their own colonies, and wasps are kinder to their families than they are to strangers. For example, scientists have found that wasps whose colonies are opened to strangers will act with the most civility toward the wasps with recognizable "family" pheromones.

The Marvels of Ants

To say pheromones direct the lives of the estimated 8,800 species of ants that roam the globe is an understatement. Rather, pheromones dictate exactly what each and every member of an ant colony will do, how it will behave, and when. Pheromones tell ants when to search for food, if the colony is being invaded, and when another ant has died. Ants have twenty-five glands in their bodies that all together produce two dozen different pheromones.

In the world of the ant, pheromones are like molecular stand-at-attention orders that each member of the family is genetically hard-wired to follow. To an ant, not responding appropriately to a pheromonal signal would seem as foreign as not raiding an open picnic basket.

Edward O. Wilson of Harvard University, who pioneered studies into chemical communication among ants, holds the exotic title of myrmecologist, which is a biologist who specializes in the study of ants. One of Wilson's most significant findings is that certain ants are capable of laying distinct pheromone "smelling" trails that tell the others where food has been found. The others can read the instructions provided in the trail and then follow it to the food.

Despite the ant's pheromone mastery, its code can be broken by some invertebrates that can unscramble the messages and edit them to their advantage. Assassin bugs, for example, are brilliant decoders. They find the ants' odor-pheromone trail and replicate it. When a stream of ants picks up on and then follows this fake trail, the assassin bugs attack the ants and eat them.

Animals: The Great Communicators

Can pheromones induce camaraderie or emotions in animals as they are thought to do in humans? When we talk of animals having emotions, we often speak in human terms. We say, "That dog looks sad," but what leads us to that conclusion? Is the dog withdrawn and droopy (as sad humans tend to be)? Does it lack energy or enthusiasm? We know how *we* feel when we're sad, but can we apply those characteristics to animals?

Higher-level expression among animals has been debated for many centuries. Aristotle believed that animals could exhibit a broad range of passions, and that humans and animals share some traits, including gentleness, fierceness, mildness, courage, and timidity. Aristotle made one very clear distinction, however: In his view, animals display their passions instinctively; that is, they act automatically, without intention.

Aristotle also theorized that animals lack virtues and vices and that only humans are capable of moral or immoral behavior. He concluded that passions are the result of instinct, whereas emotions require the involvement of the intellect. Because animals lack intellect, they can display only certain passions, not emotions.

We have seen how animals are compelled to display a full range of passions—lust, anger, sexual jealousy—as they respond to the chemical codes of pheromones. But is there more to animal behavior than the obvious motivations that are determined by pheromonal stimuli? Do animals have the ability to communicate deeply felt feelings?

Writer Jeffrey Moussaieff Masson touches on the controversial issue of romantic love among animals in *When Elephants Weep:*

The Emotional Lives of Animals. He says, "Whether it is called emotion or drive, in most scientific circles it is forbidden to say that animals love."

Masson goes on to say, "Perhaps love, the emotion, has evolutionary value." Indeed, it seems reasonable to suppose that the power of love could very well be the motivation behind ferociously protecting a nest or mating for life. But can love among animals be reduced simply to no more than a clever evolutionary method of ensuring the survival of the gene pool?

Masson tells the story of a mother elephant who went through heroic efforts to save her drowning calf. Oblivious to her own needs, the mother risked her life to save her baby. Fortunately, both survived, even after the calf found itself stranded on a ledge and the mother was swept some distance downstream in the rush of the water. The mother's face seemed awash in emotion. She was frantic over seeing her baby in danger, and she reacted much like a human mother would if her child were in jeopardy.

There are accounts of animals who have "fallen in love" with each other to such an extent that when one dies, the other retreats into melancholy and depression. One example Masson gives, as told by animal behaviorist Konrad Lorenz, involves two geese who had mated and bonded. When the female was killed tragically, the male responded with a grieving process that was strikingly similar to a human's. He had no gumption, no appetite for life. He sat slumped over, his eyes dull. And yet, after some time spent in mourning, the bird recovered sufficiently to start a relationship with another goose.

Stories exist of animals who express strong preferences for others of their species. Humans don't hand over their hearts to just anyone. We size up our potential mates: how he smells or dresses, what college she attended or where she works. However, beneath this analysis the pheromonally directed dance of attraction goes on. If a couple's pheromones are not compatible, chances are the couple won't be happy together. Perhaps it's the same for animals.

Masson writes of a male cockatoo who rejected a beautiful female bird brought to him for companionship. Perplexed, the bird's keeper introduced another female, this one in poor shape, with patchy plumage and dry, wrinkled skin. This time, the male was smitten: "The two birds immediately paired off and began rearing a series of baby cockatoos," Masson says.

Could it be that while animals rely on pheromones to instigate such deeply primitive behaviors as mating, aggression, and dominance, they are also capable of feeling and expressing emotions through chemical communication? A human mother bonds with her baby through a pheromonal connection even before the infant is born. This bond translates into feelings of intense, unconditional love for her child before, during, and after its birth. Perhaps unconditional love of this nature is to be found in the animal world, too.

Questions such as these may one day be answered through pheromone research. Until that time, we can ponder the mysteries of how animals find each other, stay with each other, even fall in love and stay committed. In his poem "The Elephant Is Slow to Mate," D. H. Lawrence describes the creature's mating rituals with words often reserved for intimate encounters between people: "So slowly the great hot elephant hearts / grow full of desire, / and the great beasts mate in secret at last, / hiding their fire."

Pets and People

Think for a moment about a special family pet, the one who day by day etched its personality into your heart. Were you and your pet communicating in ways that go beyond the obvious pat on the head or the happy wag of a tail?

The intimate connection between animals and their owners has become a popular topic. Some people claim they "speak" to their pets on a subconscious or extrasensory level. It has been shown that the presence of an animal to stroke or talk to can reduce a person's blood pressure and instill a sense of calm and well-being. Physician and author Larry Dossey writes in an article in *Body Mind Spirit* magazine: "Devotion to a pet, like a devotion to prayer, can bring about improvements in human nature, as seen in the dynamics of families. . . . Being around pets, like praying, brings out compassionate behavior in people."

Pets seem to know what we're feeling, too. Scientists have discovered that a dog's highly refined sense of smell allows it to detect human emotions. A dog might avoid the company of an angry person or retreat from a threatening posture. In the popular Walt Disney movie *101 Dalmatians,* a character says of the tale's villainess, Cruella De Vil, "The dogs never liked her. Dogs have a sixth sense about that. They can smell ill intentions."

Dogs also appear to be able to single out individuals with psychoses through their senses. For example, children with autism and other psychiatric disorders literally repel dogs with their breath. Dogs participating in such studies will veer away from the "abnormal" children, preferring to play with the healthy ones. What is it that the perceptive, odor-sensitive dog detects in the presence of someone with a psychiatric disorder? Does the animal sense something subliminal—a pheromone out of whack, perhaps?

People and animals have been connected for centuries. Beasts of burden made possible transport and agriculture and were a valued food source. The citizens of Mesopotamia bred sheep, the ancient Greeks and the Japanese kept dogs, the Norse shamans thought of their reindeer as spiritual companions, and swans were popular pets in tenth-century Britain. Roger Caras, the president of the American Society for the Prevention of Cruelty to Animals (ASPCA), believes that our transition from a hunter-gatherer way of life in the Stone Age to having pets in our homes is the result of intimate relationships with animals that occurred while they were still viewed as a means of convenience. Is it just the physical company of pets we seek or do we unknowingly communicate with them on levels triggered by chemistry?

Current research shows that pheromones are species-specific, but could it be possible that Fido and his master engage in some form of chemical communication? Perhaps someday a study will address, and possibly answer, this question.

The Nose Knows: The Human Vomeronasal Organ

When we smell another's body, it is that body itself that we are breathing in through our mouth and nose, that we possess instantly, as it were in its most secret substance, its very nature.

—Jean-Paul Sartre

While researchers have long explored how animals and insects use pheromones to "talk" to each other, the science of pheromones in humans is relatively young and has focused on a tiny but important piece of human anatomy: the vomeronasal organ (VNO).

The Erotic Nose

The story of Barbara, an anthropologist in her late thirties, and Michael, an engineer, could have been taken straight from the pages of a romance novel.

They met at a large metropolitan museum. Barbara was admiring the sensuous curves of a Rodin sculpture when Michael walked up to the exhibit. He didn't take much notice of the woman standing next to him. But, several minutes later, something changed. Something incredible began to happen right under his nose.

Michael says he became distracted. His focus on the sculpture wandered and he found himself turning his attention to Barbara. He was standing about eighteen inches from her and had an urge to touch her. *This sounds crazy,* Michael remembers, *but something about her called out to me. I looked at her, she looked at me and, like in the movies, we fell in love.* Taking a risk, Michael started up a conversation with Barbara. To his delight, she responded enthusiastically. They retreated to the museum coffee shop, continuing to talk and get to know each other. He asked her to dinner. She accepted. More dates followed, and the relationship began to progress. Deeply in love, they married six months later, and are still together.

Loving Love

Is there any one thing more examined in the human psyche than the emotion of love? Some people are lucky in love: They find the perfect fit in a partner and tend the garden of romance for decades. Others strike out again and again in affairs of the heart. We obsess over love and we endure bad loves, relationships that damage our egos and leave our self-esteem cracked and bleeding. Nevertheless, we continue to seek love despite any emotional scarring. The goal is union, connection, and only love can feed such a hunger.

One theory about love is that it facilitates our basic desire to send our genetic codes into the future. That theory certainly has merit and may well move the tides of love. Once we get caught in the wave, we risk not being able to return to shore without having incurred some damage.

As anthropologists study the evolutionary and familial benefits of falling in love, scientists investigate how love is rooted in the physiology of the body. That's why we have welcomed the word *chemistry* into our lexicon of love—it implies that a mysterious but wonderful reaction is happening within us.

When we speak of love in terms of chemistry, we describe passionate, distracting, even frightening encounters that can send us over the moon. What can we make of love that ignites instantly? Can we trust our immediate reactions to another person? Are our bodies equipped at a physical level to help us sort through and analyze our relationships? Should we believe in what Jung called "the ultimate honesty of the body"?

When Michael and Barbara met in the museum, some unknown force sparked a mutual interest in a matter of minutes. Moreover, had it been David instead of Michael in the museum that day, the chemical reaction might not have occurred, and Barbara would have walked onto the next exhibit alone, her life unchanged.

Michael and Barbara relied on a host of sensory cues when they met. For example, he liked the deep brown of her eyes and she was attracted to the sound of his voice. But there was more. As this couple followed the siren call of their attraction, they were subconsciously tuning into important messages provided by the sixth sense. In this chapter, we will look at the physical and pheromone-receiving properties of what we will call the erotic nose.

Doorway to the Sixth Sense

We don't think twice when we *see* something. Our eyes register the colors of the ocean and send the signal to the brain. We then arrive at our conclusion: "The ocean looks gray today."

When we *hear* something, we know in an instant that our auditory centers are in gear, working to tell us whether the sound is a cello, a hummingbird, or a jackhammer. When we touch something, our touch receptors zip information to our brains and we register what our fingers have landed on—soft, hard, rough, silky, sticky.

What happens when the sixth sense detects a pheromone? Even though it's taking up residence right inside your nose, you don't *smell* the pheromone. To put it another way, you don't create an immediate inventory of the sensation in question as you do with the other five senses (he *smells* wonderful, her skin *feels* soft, I can *hear* the sound of the rain on the roof, I love the *sight* of pink roses, this peach *tastes* heavenly). Why is this so? Because pheromones affect the subconscious mind.

As we have seen, the human vomeronasal organ is the sense organ in the nose that picks up and processes pheromones released by other people. The VNO contains cells not found anywhere else in the body that give the human VNO its unique chemosensory properties. The VNO is one of our most important organs because it is the doorway to the sixth sense.

When a pheromone enters your nose, it travels to your vomeronasal organ. Once there, the pheromone latches onto special receptor cells located on the surface of the VNO. The receptors compute the chemical signal and send it to the hypothalamus, a primitive part of the brain that controls a host of basic bodily functions, including skin temperature, metabolism, hormone production, appetite, sex drive, fear, aggression, and anxiety. While the cerebral cortex is what gives us our reasoning and thinking abilities, the hypothalamus directs what we do on a subconscious level. It is also where pheromones register their effect.

As the sixth sense guides the flow of the neural rivers of the brain, it gives us information not supplied by our other five senses. If you fall in love, you may think it's because you've met someone with whom you're compatible and who has similar beliefs and values. That could be true. But falling in love—or experiencing any one of the broad range of human emotions—is heavily influenced by the sixth sense.

The sixth sense resides in the region of the brain that responds to sensory cues before the thinking brain, the cortex, can get involved. It is the commander of your emotional battalion. But, unlike nonhuman animals, humans possess a well-developed thinking brain that allows us to control the urgency of chemical cues. We are able to balance our instinctive side with our rational side.

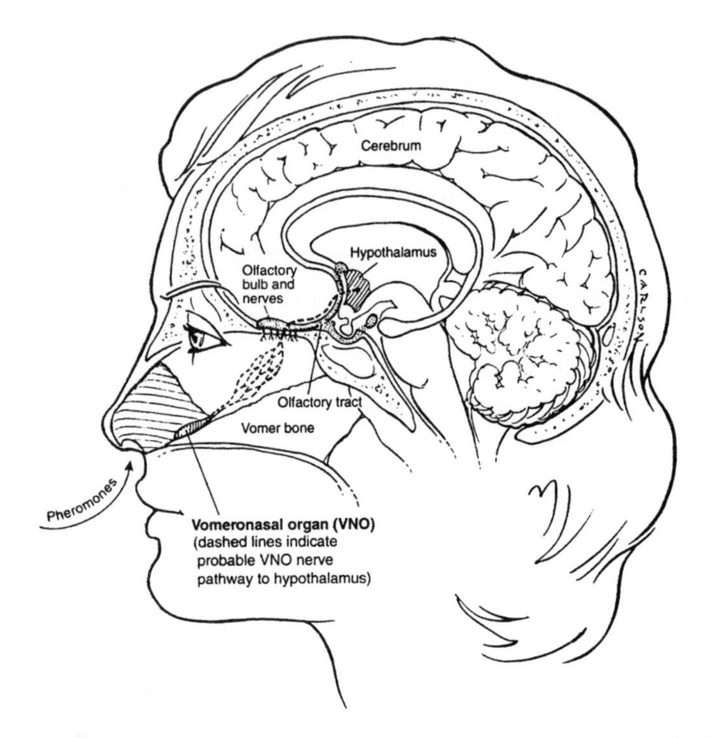

This drawing shows the position of the vomeronasal organ (VNO), which detects pheromones, in the human nose. Pheromone molecules enter the nostrils and contact the sensory receptor cells in the VNO, which in turn send nerve impulses to the hypothalamus, a key control center in the brain.

The Search for the Sixth Sense

The man whose work culminated in the discovery of the sixth sense is Dr. David L. Berliner. He looks like a kindly grandfather, except this grandfather is one of the world's top biotechnologists. Among other things, Berliner helped pioneer techniques for administering drug therapy through skin patches like the well-known nicotine and estrogen patches.

Berliner was born and raised in Mexico City, the son of parents who emigrated from Poland to Mexico during the 1920s. Berliner happened upon pheromones in the 1960s while working as an anatomy professor at the University of Utah School of Medicine in Salt Lake City. At the time, he was conducting studies into his favorite subject—the human skin (over the course of his career, Berliner has published more than 130 scientific papers on skin).

The Utah studies involved scraping shed skin cells off plaster casts that had been worn by skiers with broken limbs. The purpose of the experiment was to document which substances are found in human skin. Berliner stored extracts of the collected skin cells in a flask. But as he collected skin cells, he began to pick up on something quite unusual taking place in his lab.

The people in Dr. Berliner's lab were not a playful bunch. They did their work with little frivolity or camaraderie and went home right after work. They didn't joke with each other or even talk much. To Berliner, this was fine. It meant the work was getting done and he could focus on his experiments without distraction. But when the flask containing a certain extract from skin cells was left open, Berliner discovered, the moods of his coworkers changed dramatically. As the lab brightened with laughter and unprecedented goodwill, Berliner became more and more curious. *This is odd,* he thought. *Why are people suddenly acting nicer toward each other?*

"There was a female technician who was a hard worker but straight as an arrow," he recalls. "She came to work, did her job beautifully, didn't talk, then—good-bye! She was all business. One day she said, 'What about learning how to play cards during lunch?' Soon, we were all playing cards—even *I* was playing! The flask was open. When I put the flask back into the refrigerator, no more cards. Nobody knew why this happened. No one even imagined what the reason was. Even I didn't put it together at first."

Berliner began to wonder if his staff's mood changes just might be connected to the contents in the flask. It's important to clarify that the flask contained not ordinary skin cells but superconcentrated extracts of skin cells. An extract of skin cells, in contrast with the singular cells that slough off the body every day, is loaded with pheromones—"millions and millions" more than regular skin cells, Berliner explains.

When Berliner left the university to pursue a career in venture capitalism and biotechnology, he put the flask, which still contained the skin cell extracts, in the deep freeze and left it there until 1989, when he revisited his decades-old hunch that the extract had somehow influenced human behavior. For years, Berliner had been unable to shake his experience in the Utah lab. What kept resurfacing in his mind was what had happened when the flask was closed: The staff's behavior returned to normal. They became testy, unfriendly, and characteristically silent.

While the pheromones in Berliner's flask sat frozen, studies into how pheromones serve as chemical signals in insects and animals were gaining momentum. For years, Berliner tossed a novel thought around in his mind: *If other animals produce and respond to pheromones, why wouldn't humans?*

To test his idea that humans might communicate with pheromones and that these chemicals were most likely produced in the skin, Berliner asked University of Utah scientists Dr. Clive Jennings-White, a Cambridge University–educated organic chemist, and Dr. Louis Monti-Bloch, an M.D., Ph.D. neurophysiologist, to help him design a series of experiments. He also signed on Dr. Larry Stensaas, a University of Utah neuroanatomist.

Stensaas was charged with a difficult task. Berliner thought the flask might contain human pheromones. Still, he needed to locate the anatomical structure that could receive and process the molecules. He had to confirm the presence of a human vomeronasal organ.

Until recently, the human VNO had not received a good deal of respect in the press. In 1938 neuroanatomist Dr. Elizabeth Crosby, writing in an anatomy textbook, dismissed the VNO, saying it existed in humans during the fetal stage but didn't survive beyond that. She delivered what appeared to be the final blow when she said the VNO, only rarely seen in adult humans, was vestigial and nonfunctional. Crosby's words held ground for fifty years, until the work of Berliner and his colleagues shook her theory from its foundation.

Stensaas spent months looking at nasal specimens through an electron microscope, an instrument that focuses rays of electrons, instead of the usual light rays employed by a light microscope, to form an enlarged image of a subject. In time, he arrived at what he had been searching for: evidence of a human vomeronasal organ.

Stensaas's micrographs strengthened Berliner's hypothesis that humans have the necessary anatomy to process pheromones. They revealed that the human VNO appears as a pitlike opening in the nasal wall, beyond which lies the organ itself, a tube lined by tall, narrow cells—the pseudostratified columnar epithelium.

Meanwhile, Dr. David Moran, a cell biologist and electron microscopist, and Dr. Bruce Jafek, an ear, nose, and throat specialist, had independently examined the human VNO and found it to be present in all of the patients they had studied. As they worked together, Moran and Jafek, who teamed up to study the human sense of smell in 1981, began to discuss the accessory olfactory system (or, in simpler terms, the VNO–pheromone system) in mammals. "It occurred to both Bruce and me that we're mammals. Being mammals, it would be strange if humans didn't have a functioning vomeronasal organ," says Moran. "And just as the human sense of smell is reduced in size, so may this accessory olfactory system be reduced in size—but it really ought to be there."

The scientists began to look for the organ in the noses of patients who came into the clinic at University Hospital in Denver. Using a sophisticated microscope made for microsurgery, they found the VNO in all the patients they examined.

From there, Moran and Jafek embarked on a more formal investigation of the human vomeronasal system. They were interested in three things: The frequency of the human VNO (does everyone have one?); the ultrastructure of the organ (what you can't see with the naked eye); and whether the VNO contains sensory receptor cells. Without special receptor cells onto which pheromone molecules can latch, the VNO would not be able to send nerve impulse signals to the brain.

The scientific world was poised to do battle against anyone bold enough to claim that the VNO existed as a functioning organ in humans. In many circles, the idea of humans communicating with pheromones was considered to be preposterous, and anyone speaking of chemical communication among humans was thought to be off his scientific rocker.

In animals, the VNO transmits pheromonal messages via the vomeronasal nerve, which terminates in the region of the animal's brain called the accessory olfactory bulb (AOB). But, researchers had not been able to locate the vomeronasal nerve or the accessory olfactory bulb in people.

One reason the AOB might not be visible in adults is because of the way the brain develops. Anatomy texts have traditionally stated that the AOB "involutes" during the second trimester and disappears by the time the baby is born. Moran and his colleagues had located a well-developed AOB in fetal brain specimens at the second and third month of gestation, and some of the fetuses had well-defined bulbs as late as four to five months into gestation. None of the brains showed a clearly defined AOB after the five-month stage.

Here we have a brain structure that is not visible after the fifth month of gestation, but does that mean it doesn't exist in adults? Moran isn't convinced, although to date the AOB in adult humans has not been located. However, Larry Stensaas, continuing his studies at the University of Utah, has found no evidence of any degenerative processes (neurodegeneration) that might affect the AOB during the fetal stage. Thus, there is no clear-cut reason for the AOB to disappear.

But here is one theory: As it develops, the human brain undergoes massive changes, particularly in the frontal lobe. The frontal lobe is so large that it exerts extreme pressure on other areas of the brain, including the AOB. As the frontal lobe grows, essentially elbowing its way into the confines of the skull, it could cause the AOB to stretch into a flat band of cells. Thus, the

AOB's appearance in adults might not resemble a bulb at all, but a rubber band pulled taut.

In their follow-up VNO studies, Moran and Jafek carefully examined the noses of two hundred volunteers. Peering inside with special headlamps and magnification devices, they were able to locate the VNO in every subject. Moran then used an electron microscope to probe even deeper into surgical specimens of nasal tissue.

This time, there was no question: Moran and Jafek were looking at the human vomeronasal organ. The old school of science had stated that the human VNO rarely existed in adults. Moran and Jafek had proved otherwise. "We couldn't believe it when we saw the VNO in every instance," says Moran. "We found that there is not only an opening, but an organ behind it. And, we found that there are some very interesting-looking cells in it. *And, it looked like a sense organ.*"

The human VNO system is made up of two tiny organs that sit deep inside both nostrils. Although in some people you can see the VNO opening, or pit, with the naked eye, in most cases the pit must be magnified to be visible. That may explain why the VNO had previously only been observed on rare occasions.

With the presence of the human VNO confirmed, the task turned toward finding the link between the organ and the brain, the freeway on which pheromonal messages can travel. That freeway is the vomeronasal nerve.

Larry Stensaas and his colleagues looked again to the fetal brain for answers. Earlier research had concluded that the fetus has a vomeronasal nerve that connects the VNO to the accessory olfactory bulb. However, the nerve was thought to exist only during the early months of gestation. Before moving on to other ventures, Stensaas investigated the probability that the VNO maintains its link to the hypothalamus into adulthood. Subsequent studies have indeed shown this to be the case.

Moran and Jafek also discovered that the VNO lies separate from the olfactory epithelium (the cells that make up the olfactory system, which is responsible for the sense of smell) inside the nasal cavity. Eventually, they realized that while the two systems are separate and work independently, they run parallel in the following way:

VNO receptor → vomeronasal nerve → accessory olfactory bulb of the brain
Olfactory receptor → olfactory nerve → olfactory bulb of the brain

When Berliner found out about Moran's micrographs, he contacted him with an irresistible proposal: Come work with me. Moran accepted, joining the team of respected scientists that Berliner had assembled for his journey to the sixth sense.

The pieces of the pheromone puzzle were slowly shifting into place. David Berliner and his fellow scientists now sought to find out if the human VNO—which existed beyond a doubt—responded to pheromones. Berliner would turn to the contents in his flask, his confidence boosted by David Moran's elegant micrographs of the VNO.

Pheromones and the VNO: Solving the Puzzle

A neurophysiologist whose accent is flowered with the influence of his native Uruguay, Dr. Louis Monti-Bloch works in the psychiatry department of the University of Utah in Salt Lake City. Monti-Bloch, whose pioneer investigations into the workings of the human VNO have provided us with much of the framework for our study of the sixth sense, designed and conducted a series of double-blind studies to determine how the VNO reacts to pheromones.

To perform the experiments, Monti-Bloch developed a mini-probe, a device that delivers pheromones and other control substances directly to the VNO and records resulting electrical activity on a computer. The mini-probe detects subtle changes in the surface voltage of VNO cells and illustrates that activity with the help of an electrovomerogram (EVG).

Monti-Bloch needed to devise a method of delivering the

pheromones that would prevent them from leaking into the adjacent olfactory system; should this happen, the scientists would be unable to obtain accurate VNO test results. He surrounded a hair-thin wire electrode with two concentric plastic coverings that ensured the substances inside would be delivered where they were supposed to, and not wander to nearby olfactory cells.

The device worked beautifully. It delivered test chemicals to the VNO in a micropuff of humidified air mixed with pheromones or non-pheromone control substances. When placed directly on the VNO, the electrode picked up any electrical activity caused by the chemical inputs. It could also be trained on the surrounding respiratory tissue or the olfactory epithelium so that comparisons of electrical activity could be obtained from several other sensory and non-sensory locations inside the nose.

Monti-Bloch could also monitor what the volunteers experienced during the tests. The subjects were awake during all phases of the experiments, so they were able to describe how their feelings and moods changed (or didn't) when certain substances were delivered into their noses.

Monti-Bloch's test substances were pheromones synthesized in a lab by Clive Jennings-White, his colleague at the University of Utah. Jennings-White had devised a way to duplicate the molecular structures of pheromones derived from the skin cell extracts in Berliner's now-famous flask.

One of the keys to the VNO-functionality experiments was differentiating between the sense of smell and the VNO. If the scientists were indeed poised to reveal the presence of a sixth sense, they had to prove that this pheromone-sensing system acted independently of the sense of smell. To do this, they selected a series of strong-smelling olfactory stimulants, including clove oil and cineole. Monti-Bloch then signed up forty-nine volunteers ranging in age from eighteen to fifty-five, all of whom seemed more than willing to flare their nostrils for science. He outfitted them with electrodes for monitoring galvanic skin response, brainwaves, respiration, and heart rate. He also performed psychometric tests to explore behavior.

While the volunteers reclined comfortably on exam tables, Monti-Bloch delivered tiny puffs of pheromones into their noses. In all cases, the VNO responded immediately, sending a rapid-fire staccato of electrical activity through the mini-probe and into the computer. The olfactory cells, meanwhile, showed no response to the odorless pheromones.

Continuing with his experiments, Monti-Bloch delivered puffs of olfactory stimulants into the noses of the volunteers. The olfactory cells zipped into life . . . and the VNO sat quiet. This was clear evidence that the VNO and the olfactory cells do not react to the same substances. The VNO requires a pheromone to produce electrical activity, while the sense of smell requires a scent molecule to perform its work.

What's remarkable about this experiment is that the VNO responded to *picogram* quantities of pheromones. A picogram is a millionth of a millionth of a gram—a staggeringly small quantity.

In general, pheromones are *sexually dimorphic*—some are more active in males and others are more active in females. Monti-Bloch found this to be the case with his test pheromones; some had a more pronounced effect on women while others had a more pronounced effect on men. Specifically, the synthesized pheromone ER-670 (a version of the naturally occurring human pheromone androstadienone) shows more activity in women and ER-830 (a version of the naturally occurring human pheromone estratetraenol) shows more activity in men. To put it more simply, women react strongly to male pheromones and men react strongly to female pheromones.

In a later experiment, Monti-Bloch delivered puffs of odorless human pheromones from Berliner's flask onto single VNO cells that had been harvested from research volunteers and cultivated in a petri dish. The result stunned him: The single VNO cells fired in reaction to the pheromones but showed no response to odors. This study showed that the VNO contains neurons that jump to life in the presence of a pheromonal stimulant, even when separated from their homebase VNO system.

While early VNO experiments had used synthesized human pheromones, later studies incorporated the use of *vomeropherins,* which are a second generation of molecules designed to act on the VNO but are not produced naturally in the human skin. (*Vomero* refers to "vomeronasal organ" and *pherin* means "to convey or deliver.") A vomeropherin is a molecule that has a physiological effect on the VNO. It has been produced in a lab and can be altered to elicit a response from the VNO and the hypothalamus.

As of this writing, Clive Jennings-White and his colleagues have created more than 1,000 vomeropherins. Working under the umbrella of Pherin Pharmaceuticals, Inc., a biotechnology firm that is headed by David Berliner and based in Menlo Park, California, Jennings-White and his staff have already received and are applying for additional patents to protect their synthesized vomeropherins. Pherin Pharmaceuticals is working with the U.S. Food and Drug Administration to develop pharmaceutical vomeropherins for the treatment of a number of human ailments.

While human pheromones are preprogrammed to do what nature tells them to do, synthetic vomeropherins can be tailored to produce specific effects on the VNO and in the brain. Synthetic vomeropherins are many times more active in the human VNO than are pheromones. To date, Berliner's team of scientists has conducted numerous vomeropherin experiments on hundreds of volunteers. The results have been consistent.

Piping vomeropherins into the VNO causes dramatic electrical activity in the hypothalamus, the part of the brain that controls human physiology. The response varies with the type of vomeropherin applied. One slows down a person's heart rate while another speeds it up. A different type changes skin temperature and affects the size of the pupils. Another causes significant changes in muscle tension.

Some study volunteers report feeling instantly relaxed when

a certain vomeropherin hits their VNO receptors. This relaxation is manifested in decreased respiratory rates; the vomeropherin in question slows the body down.

One vomeropherin increases alpha waves (the brain pattern associated with relaxed states), while another increases beta waves, which make a person feel alert. Some vomeropherins even put people in better moods and reduce their feelings of negativity and anxiety. Others are capable of altering hormone levels in the bloodstream—testosterone in men and estrogen in women, for example. This is further evidence that the hypothalamus is connected to the VNO, because the hypothalamus regulates the body's hormonal systems.

Louis Monti-Bloch, who believes the human VNO may be at least as sensitive as the sense of smell, recalls his own experience with synthesized human pheromones and vomeropherins: "When I started doing this work and was preparing the substances in the lab, in a very subtle fashion I began to notice changes in myself. I felt different. I was working very late at night—it was past midnight. So, I went home and got up very early the next morning and I felt great! Later, when I was back in the lab, I found I could produce feelings of alertness in myself [in the presence of the substances].

"Then I exposed myself to the substances intentionally. I felt very nice. I had decreased respirations and was in a state of calm. My muscles also relaxed." Monti-Bloch and his colleagues presented their findings at an international symposium on human pheromones in Paris in 1991 and published their work in the October 1991 issue of the prestigious *Journal of Steroid Biochemistry and Molecular Biology.*

"I think of the brain as a complex structure of millions of neurons and neural connections," says Monti-Bloch of the mysteries of chemical communication and the human brain. "The brain has many inputs from sensory organs that are analyzed and processed to output signals. We have to interpret the human VNO as an important sensory organ in the context of all the other sensory organs. We are constantly releasing pheromones and receiving chemosensory inputs we're not even aware of. This is very real."

Trust Your Nose

Go to a mirror and look at your nose. Really study it, view it from all angles. Do you have a new appreciation for it, based on your knowledge of the vomeronasal organ it contains and the pheromones it processes and sends to your hypothalamus?

The next time you are in a social situation, spend some time thinking about your nose and the noses of the people around you. You may be engaged in a sparkling conversation about the state of world affairs—but wait a minute. The real conversation, the one that strikes in the primal depths of your brain, is the unspoken chemical conversation you are having with your friend.

The phrase "the nose knows" certainly applies to our discussion of human pheromones. Your nose knows when a person makes you feel uncomfortable, sexy, desirous, or friendly. Your nose knows which situations and people leave you feeling hot or cold. No longer just the seat of the sense of smell, the nose—the erotic nose, the discerning nose, the judgmental nose, the romantic nose—processes the chemical signals that float in the air around us.

There may be something a bit discomfiting about what's going on in our noses all day, every day. Perhaps our noses are the ones in control. They have a direct line to our emotional brains and they make decisions for us, whether we like them or not. Of course, we can ignore those subtle messages—start a friendship with someone we know we should avoid, for example. But given the experiences of a number of people we interviewed, blocking out the warnings of the sixth sense can sometimes lead to trouble.

Janet, a freelance writer, told us about her attempted, and failed, friendship with another woman. From the start, Janet says, her interactions with this woman were strained and awkward. "I couldn't figure it out," Janet explains. "It wasn't as if either of us was doing anything wrong. We just couldn't click. Maybe it's more accurate to say I didn't click with *her*. Something kept nagging at me, but I ignored it. I knew later that my gut feelings had been telling me, ever so quietly, that I couldn't trust this person.

"Well, that's exactly what happened. I couldn't trust her. Without going into details, let's just say she really let me down. I knew from the beginning that my sixth sense was telling me to be careful, but I ignored it and I ended up paying the price. It's funny, but when I met her boyfriend, I had the same reaction to him! My husband experienced a similar reaction, too. Maybe this woman and her boyfriend had pheromones that were perfect for each other—but certainly not for me!"

This is an important lesson: To allow the sixth sense to perform its work unimpeded, we must ease our grip on controlling the situations and events in our lives. If we heed the call of the sixth sense, we give in to pure chemical communication. For some of us, this is not a pleasant thought. We revel in control, and in the realm of the sixth sense, control is forced to take a backseat.

"The VNO is the sense that operates at the level of the subconscious," says Dr. David Moran. "I think that if you look at human behavior, the subconscious runs the show. And people don't like that. They don't want to admit it. People tend to be control freaks, and they like to think they can consciously control everything they do."

Do his kisses leave you feeling weak? It may be his pheromones. Is she the woman of your dreams? Maybe it's a pheromonal thing. Consider for a moment that you may have met the *pheromones* of your dreams. Don't ignore your sixth sense. It can be a trustworthy guide through the complex world of human interaction.

The Brain Dance of the Senses

When the eyes see, the ears hear, and the nose breathes, they report to the heart. It is the heart that brings forth every issue, and the tongue that repeats the thoughts of the heart.
—Ptahhotep, Egyptian vizier, c. 2350 B.C.

We asked couples what had attracted them to each other initially. Was it the color of her eyes that drew you in? Did you like the sound of his voice? Did she touch you on the knee, ever so slightly, and send an electrical charge through your body? Did his jokes make you laugh?

One couple's story stands out as a good example of what can happen chemically when two people meet. The introduction of Peter, a well-known musician, and Jessica, a playwright, took place on an airplane. Peter recalls seeing Jessica across several aisles of seats. "The color of her hair spoke to me instantly," he said. "It's a lovely golden color. I couldn't stop staring at her."

Peter noticed that the seat next to Jessica was empty. He took a deep breath and approached her. "I felt like a complete idiot, standing there in the middle of the plane, asking this beautiful woman if I could join her. She must have trusted me because she said yes."

Jessica agrees. "I recognized him, of course, because I was familiar with his music. For some reason, I did trust Peter, and this was aside from seeing a face I'd seen before, although never in person. He has a nice face, an open face. Maybe that's why I didn't turn him away."

Once settled in next to Jessica, Peter's eyes moved away from her hair, the physical trait he had seen first, and he began to fall in love with the sound of her voice. He then found himself reacting to her perfume—but was it her perfume, or her own scent? At one point in their conversation, Peter reached over and touched Jessica's hand briefly. He noticed that her skin was soft and smooth.

Jessica, too, began to pay attention to more and more of Peter's attributes. She found herself drawn to him physically. He smelled nice. She liked his laugh. She said she felt safe in his presence.

The flight ended, and both Peter and Jessica were disappointed at the prospect of parting. They exchanged phone numbers and promised to keep in touch. Four years later, they are a happy, energetic couple—and very much in love. Even their busy lives, which take them around the world and to their several homes, don't get in the way of their chemical attraction, an attraction that is evident in how they look at each other, and how each responds to the other.

It's clear that Peter and Jessica relied on a complex interplay of sensory cues to discover their mutual attraction layer by layer. They did have a number of things in common, and those similarities helped propel their first meeting into a courtship and an ensuing relationship. Still, their story reveals the degree to which their senses were involved during their first exchange aboard the airplane.

Humans are sensory creatures. Every second of our waking hours is spent sifting through the myriad stimuli that filter into our sense organs and travel to our brains for processing and decoding. When we meet someone, our senses jump to attention and begin to deliver information that helps us determine whether the person standing in front of us is appealing.

But how does it all happen, and in what order? That question still challenges scientists researching sensory systems. The order in which sensory cues are processed depends on the parameters of each individual situation. So, rather than attempt to assign an easy-to-understand formula to each sense, it's better to view them as separate biological systems, each vitally important to the final result: sensory perception.

The ancient Greeks were the first to divide the human senses into the five categories of touch, sight, hearing, taste, and smell.

The sixth sense has not been named officially, but it might be accurate to call it the *pheromone sense*: a system of chemical communication that processes those airborne molecules that facilitate subconscious communication between people. When an invisible, odorless pheromone molecule enters the human nose, it encounters the vomeronasal organ. The pheromone and the VNO then begin to communicate via a succession of chemosensory signals. Once stimulated by pheromones, the human VNO responds by sending messages down the highway of neurons that terminates in the hypothalamus.

While the brain is the processing and distribution center for all the senses, the pheromone sense and the sense of smell are the only ones connected directly to the oldest region of the brain, the region that was in place long before the seat of consciousness—the cerebral cortex—evolved to its present size.

Given that the pheromone sense is processed in the hypothalamus, can we surmise that humans live by feelings and behaviors evoked by pheromonal stimuli? While this is certainly what happens in many animals, humans are somewhat different. Humans do indeed process pheromonal cues, but for the most part we don't follow those cues with blinders on and all logic thrown to the wind. We take in and process pheromones, and the hypothalamus tells us what to do at a subconscious level. Whether we choose to listen to the call of the hypothalamus, however, is largely dictated by the degree to which our other sensory and cognitive systems interject their own information. When you take in another person's pheromones, your cortex may not speak up until the pheromonal signal has gone first to your VNO and then to your hypothalamus. Only then can your thinking brain intervene and help guide your judgments and decisions about other people.

To illustrate, let's suppose that a man and a woman meet for the first time at a party. Because they are standing close to each other, they will unknowingly exchange pheromones. The pheromones will travel to the chemosensory receptor sites on the VNOs of both the man and the woman, thereby sending chemical (pheromonal) information to their brains. This information, processed subconsciously, registers as the immediate response—a crucial part of the often-talked-about first impression. If a man and a woman each experience mutually positive first impressions, their feelings—the figurative "green light" of the situation—will foster continued conversation and, perhaps, expressions of mutual interest. The man and the woman will proceed to learn more about each other.

As their interaction advances past the introductory stage, their other senses will become more involved. The man may love the woman's pheromones but be turned off by the way she dresses, or, in other words, by what his eyes tell him. The woman may also love the man's pheromones but be unimpressed by the tone of his voice, an impression transmitted through her ears. What this means is that pheromones and the sixth sense will not guarantee that a relationship or a friendship will survive the input of the other senses. What pheromones do tell us is whether the stranger with whom we are shaking hands is at "first sniff" acceptable to us on a subconscious level.

Nevertheless, we all have heard stories about couples who claim to have fallen in love within minutes of meeting. We define this as love at first sight, but maybe it's more appropriate to call it overall sensory approval at first sight. When attraction meshes on all sensory levels, it's as if we are being told, sometimes rather blatantly, that our senses have come together to applaud the person in question. How might across-the-board sensory approval occur? If the people are in physical proximity, the first step could very well be pheromonal communication, in which the sixth sense dispatches its signals to the hypothalamus via the VNO. The other senses would then follow suit. (*She is beautiful and smells wonderful. He has a deep, sexy voice and strong arms.*)

We cannot say exactly how the six senses participate in a first meeting or in which order they offer up comment. Perhaps the sense of smell is one of the first to register, telling us whether the scent of the person standing in front of us is appealing or repugnant. Maybe visual cues strike first, forcing us to zero in on the way the person looks. Maybe a touch from the other person—a quick tap on the arm or a brush of fingers against the wrist—feels surprisingly wonderful and lays the foundation for further sensory appreciation. However, some scientists studying human pheromones believe the sixth sense is the one that kicks in first, even before the mercurial sense of smell, when two people are face-to-face. Says pheromone researcher Louis Monti-Bloch, "Underlying all your impressions of the person is your swift and subconscious impression of his or her pheromones."

Sensory Navigation

Humans are lucky. We possess more than enough sensory machinery to guide us through life. Eighteenth-century French philosopher Étienne Bonnot de Condillac, writing in *Traité des sensations,* stated that the human ability of cognition is related directly to our sensory setup and the messages the senses send to the brain. Without our senses, he believed, we could not perform even the simplest of tasks and would be unable to reason at the most basic level. We empathize with people who lose one of their senses because we can imagine the deprivation that would result. How could we possibly navigate through the world without all our senses intact?

Aristotle believed that touch was the one sense essential to an animal's survival. He wrote in the fourth century B.C.: "If an animal is to survive, its body must have tactile sensation . . . it is clear that without touch it is impossible for an animal to exist." You might disagree with Aristotle. To you, another sense might be more important. People engage in conversations about which sense they could most easily live without, which sense they believe is expendable, not necessary for survival. Some say they could get by without their hearing, while others say they could exist without their tactile abilities or sense of taste. But, people rarely discuss what might happen if the sixth sense were the one to be lost. What if you couldn't process pheromonal signals from other people? How would this sensory deficit affect your relationships and your decisions? People born without a properly functioning sense of smell also lack a VNO. Such people, who have the congenital condition called Kallmann's syndrome (described in more detail in chapter 6), are slow to develop sexually and may never pursue or express interest in sexual relationships. Researchers at the Center for Sensory Disorders at Georgetown University found that approximately 25 percent of people with smell disorders also have an impaired sex drive, and laboratory studies on rats have found that animals whose olfactory nerves are severed cease to engage in mating behaviors. This is an important point to consider. Lacking a VNO and a sense of smell appears to have deleterious effects on a person's sexual maturation and reproductive ability.

We must be careful here to point out that the sixth sense does not by design usurp the significance or workings of the other five senses in humans; we propose that it works in concert with them. But, this is not the case for many other mammals. For example, a rodent's response to pheromonal signals from its species occurs without the mitigating effects of other sensory input. As a result, the rodent is pheromonally driven in certain aspects of its life—mating, territory marking, dominance, and aggression.

Still, humans *are* pheromonally driven when it comes to making decisions about other people. You don't know exactly when your VNO awakens to pheromonal signals, but your primitive brain registers these experiences immediately and is highly skilled at tracking and documenting them.

Our Miraculous Gray Matter

Inside the bony protection of the skull lies a remarkable example of evolutionary achievement. The human brain, approximately 85 percent water and weighing just three to four pounds, controls every function of the human body, from the tiniest cell division to the most complicated physical maneuver or intellectual thought.

For Aristotle, the brain was responsible for cooling the body and regulating its temperature; however, he believed that the seat of human thought rested not in the gray matter but in the heart. Modern research has proved that the brain does far more than regulate bodily processes; it is the guiding force of the body, the keeper and instigator of our thoughts and emotions. Surgeons can replace someone's heart, liver, or bone marrow, but they can't replace a brain that has begun to die.

Despite its formidable powers and ability to store trillions of bits of information, a brain on a table resembles a rather unimpressive glop of jelly. At full development the human brain contains a trillion nerve cells, 100 billion of which talk to each other constantly via neurotransmitters, or neurochemicals. Neurochemicals move from neuron to neuron along pathways called axons and by way of branchlike dendrites in the complex play of human action and thought. Thanks to its abundance of nerves, the brain is able to take in information sent by the body and from the world and process it, as well as send its own instructions: speed up heartbeat, signal hunger, slow metabolism, alter the flow of hormones.

The intricacy of our neurochemical makeup, and how these chemicals work with our neurons to generate activity in our brains, is daunting. Notes Susan A. Greenfield in *The Human*

Mind Explained: An Owner's Guide to the Mysteries of the Mind: "Each neuron can have tens of thousands of links with other neurons. The number of possible routes for nerve signals through this vast maze defies contemplation."

The main parts of the brain are the brain stem, the cerebellum (responsible for coordinating movement), and the cerebrum. The brain is divided into two bilaterally symmetric halves, or hemispheres, which are conjoined by the corpus callosum. Each half contains four lobes and is covered with the wrinkled external layer of the cerebral cortex, which is about 2 millimeters thick and about 1½ meters in total surface area.

The brain stem is one of the oldest parts of the brain, the region that developed and grew before the overlying cortex (cortex is derived from the Latin word meaning bark). Because it is the seat of our autonomic bodily functions and the originating point for spinal nerves, the brain stem is truly the core of the central nervous system. It gives us our basic abilities and directs us at a subconscious level.

The medulla oblongata extends from the spinal cord in a kind of bulb formation that sits at the base of the skull. It regulates the activity of the blood vessels and the heart, as well as respiration and other autonomic activities. Other important brain structures are the pituitary gland, a small, oval endocrine gland attached to the base of the brain and whose secretions help control the other endocrine glands and influence growth, metabolism, and maturation, and the hippocampus, which has a central role in memory.

With the exception of the sense of smell, the senses are regulated by the thalamus. The thalamus acts as a processing center for the millions of messages, transmitted via nerve signals delivered by the sense organs, that bombard the brain constantly. Once sensory signals have passed through the thalamus, they are routed to the cerebral cortex, which taps into our consciousness and allows us to give the sensory nerve impulse a label—I see a sunset, I taste a banana. Smell, one of the most primitive of the six senses (as primitive, we suppose, as the sixth sense), sends its signals directly to the subconscious part of the brain, as well as to other areas of the brain responsible for storing and evoking emotion and memory.

The hypothalamus, which sits just beneath the thalamus and above the pituitary gland, makes up only 0.3 percent of the total weight of the brain, but is the structure that controls our basic human drives.

How many times during the day do you refer casually to your sensory systems? "I'm touched," you say to a friend who has delivered a bouquet of flowers to your door. "I see . . ." you say to someone who is trying to explain something to you. "I hear through the grapevine that you're leaving town," you say to your neighbor.

The language of our physical sensory systems is firmly embedded in our everyday speech. When we say we're touched by an act of love, we aren't really being *touched* by that act but are affected at an emotional level. In our efforts to verbalize how we feel about the kindness of a friend, we search for ways to express our deepest emotions. What better way to do so than by referring directly to the senses?

Our brains are divided into specialized segments for processing the vast quantity of information perceived by our sense organs. If our eyes capture the last twinkling light of dusk, that information, in the form of light rays, is sent along the optic nerve to the optic lobe, which gives us our ability to see and tells us what it is we are looking at. If we hear a rousing symphony, our ears take in the vibrations of the music and send analogs of them to the auditory cortex, where we register the numerous instruments involved in blending the notes. When we detect the odor of chocolate cake, our olfactory cells drive nerve impulses to the olfactory bulb in the brain and then straight to the limbic system. When we are touched, those sensations move swiftly from the surface of the skin to a highly specialized sensory region of the cortex.

Although the complexity of the human brain makes it impossible for us to explain its workings in detail here, it is important to understand the delicate interplay of the six senses and the brain. The senses are our windows to the world. People placed into experimental sensory deprivation, removed from the richness and color of the sensory environment, return to the real world lugging a host of psychological problems.

In the next section, we will talk briefly about the senses of touch, sight, hearing, taste, and smell. The notes at the end of the book provide references for readers who wish to learn more. We have devoted more space to the discussion of the sense of smell because pheromonal communication and olfaction are so intimately linked.

Touch

In the physiological sense, a simple touch from a lover can speak volumes. The writer Colette said, "Massage is a woman's sacred duty. Without it, how can she hope to keep a lover?" When we are touched by someone we love, our brains react by releasing the hormone oxytocin from the pituitary gland into the bloodstream. This hormonal "flood" sensitizes our bodies to pleasurable touch.

Aristotle challenged the assumption that humans have just five senses. He thought that the sense of touch might truly encompass more than one single sensory ability. In his *Treatise on the Principles of Life*, Aristotle posited that touch actually comprised a variety of sensations: rough-smooth, hot-cold, and hard-soft.

A number of things occur when the skin is touched. Our ability to detect a variety of touches, from the most delicate to the most forceful, resides in the brain's somatosensory cortex. The sensitivity of a specific body part is related directly to the size of corresponding areas within the somatosensory cortex where touch is registered. For example, the lips, tongue, and thumb all have larger "patches" in the somatosensory cortex than do the head, shoulder, and eye.

While the brain is responsible for creating definitions of touch, the skin is the vehicle by which touch occurs. The skin contains special touch receptors that deliver messages to the brain, and our most sensitive body areas can contain thousands of these touch receptors per square millimeter of skin.

Sight

You are looking at a caravan of cumulus clouds drifting over a mountain range. The sky is blue. The grass is green. A bird with a yellow tuft of feathers on its head wanders into your visual field for a few seconds and then lands on the branch of a nearby tree. You notice that the color of the shirt you are wearing has faded in the wash.

When we take in the views of our world, we do not process each image as a chunk of sensory information. Instead, everything we see is registered in the eyes as a series of light rays. Light rays hit the outer layer of the eyeball (the cornea) and then pass through the pupil, the opening in the iris that responds to fluctuations of lightness and darkness. From there, light rays travel through the eye's lens, which focuses the image. The image is then projected onto the retina, an area in the back of the eye that contains millions of light-detecting cells called rods and cones. When the rods and cones perceive an image, they send that image, in the form of nerve impulses, along the optic nerve, which terminates in the brain's primary visual cortex in the optic lobe. Once there, the image registers as a whole.

The capacity of the eye to latch onto minute amounts of light is remarkable. Researchers have found that the retina can detect and process a single photon of light. (A photon is a unit of retinal illumination.) However, human sight pales in comparison to the visual acuity of certain birds of prey. For example, an eagle's eye contains millions more densely packed rods and cones than a human eye, which allows the bird to spot animals on the ground from distances of up to three miles.

Hearing

Sound can be soothing, jarring, invigorating, irritating. Although a complicated process of brain activity must first occur before a sound is registered, sounds take less than one second to form in the brain.

What we perceive as sound—a piano's delicate upper notes, the wind moving through the leaves of a tree—is really a series of sound waves produced by an object that vibrates and sends those resonations through the air and to our ears. Once at our ears, sound waves move to the eardrum, which starts to vibrate. Vibrations originating at the eardrum then move through the bones in the middle of the ear—the stapes, the malleus, and the incus—and then into the inner ear's cochlea, which responds to the varying frequencies of sound.

The cochlea houses the basilar membrane, on which sits the organ of Corti. The organ of Corti contains thousands of hair cells that move in response to the stimulating effect of sound waves. The hair cells in turn send nerve signals along the cochlear nerve. Eventually, sound waves reach the medulla and then the auditory cortex, where the actual sound is perceived.

Taste

Given the many varieties of foods and flavors, it's hard to believe that the human tongue can register only four tastes: sweet, salty, bitter, and acidic. Four tastes might seem unnecessarily limiting to us, but they were all our evolutionary ancestors needed to determine which foods were good and which foods were bad.

The sense of taste works by identifying chemicals that form salty, sweet, bitter, and acidic flavors. The taste buds are formed by clusters of chemosensory receptor cells that take chemical inputs and send signals via the gustatory nerves first to the nucleus solitarius region of the medulla and then to the thalamus, which relays the sensation to the brain's taste centers in the cerebral cortex. Most taste buds are housed on the tongue, but some migrate toward the palate and into the throat. (We each possess ten thousand of them at any given time, and they regenerate constantly, essentially replenishing our sense of taste with new equipment.)

Interestingly, some of the nerves that connect the taste buds to the brain terminate in the limbic system, where many emotions are processed. Scientists posit that this neural connection can lend emotion and feeling to certain foods and events in which food plays a key role, such as holidays, birthdays, and weddings.

Taste and smell are intricately linked: Ninety percent of what we call taste depends upon the sense of smell. What happens when you have a cold and your nose is stuffed up? You can't taste things as well as you normally can. If you pinch your nose together and take a bit of an apple, you'll probably have a difficult time figuring out what is hitting your tongue. Most likely, rather than relying on the taste to identify the food, you'll turn instead toward its familiar texture. You may not get the full flavor of the apple, but you will recognize its coarse, grainy surface.

Smell

Marcel Proust wrote in *Remembrance of Things Past*, "After the people are dead, after the things are broken and scattered, taste and smell alone, more fragile but more enduring, more unsubstantial, more persistent, more faithful, remain poised for a long time, like souls, remembering, waiting, hoping, amid the ruins of all the rest."

Proust's eloquent rendering of the combined marvels of taste and smell is not only lovely to read, but accurate. Taste and smell are so interconnected that at times they appear to be one sense. Indeed, they are the senses that can affect us very deeply. This is because they are both connected (smell directly, taste more indirectly) to a web of neurons leading to the hippocampus, the area of the brain that houses memories and calls them forth so accurately and poignantly.

Although the human sense of smell is meek when compared to the olfactory powerhouses of many other animals, we are fortunate enough to be able to recognize approximately ten thousand odors. Helen Keller certainly must have known first-hand the significance of the sense of smell; in her world of sensory deprivation, her nose provided a door to "viewing" life. She called the sense of smell "the fallen angel of the senses," referring to the lack of attention paid to it historically.

The sense of smell, despite its low placement on the sensory totem pole, is one of the most important senses at all stages of life. Even sperm cells "sniff" their way to the egg as they compete by the frenetic millions for a chance to fertilize the ovoid goddess. Researchers at Johns Hopkins University have hypothesized that rodent sperm cells sniff for the egg by processing a series of proteins similar to the ones used by the animals to identify odors and scents.

Smells can push memories filed in the recesses of the brain unexpectedly to the forefront of emotion and experience (e.g., Proust's madeleines). A few molecules of a familiar smell are enough to transport you back to your mother's kitchen, where you're watching her lace a crust across a pie. A chemical smell might return you to the yellow bus that carried you to school with its odor of green vinyl. Another smell, a sweet cologne, perhaps, might take you back to the night you first made love to your college sweetheart. A smell can bring on a wave of unexpected emotion—happiness, sadness, wistfulness, regret. Rudyard Kipling understood the power of scent when he wrote, "Smells are surer than sights and sounds to make the heart-strings crack."

Scents are integral to many cultures. Anthropologist Margaret Mead reported that some primitive tribes will instigate wars with each other if they don't like the way their opponents smell. The Malay people use incense in their rituals because they believe the pleasant odor will travel to and win them favor with their moody, temperamental gods. A popular intellectual pursuit among the Japanese is the elaborate production of the kodo incense ceremony. A practice that began in the sixth century and became highly popular among the aristocracy of the seventeenth century, kodo challenges participants to identify the odor of numerous scented substances and woods, and to ascribe to each scent a literary theme that reflects the nature of the odor.

The sense of smell is triggered when the nose encounters a molecule of scent. When you dip the tip of your nose into the fragrant petals of a lilac, for example, airborne scent molecules sweep into your nostrils and gather at a thin sheet of tissue in the roof of the nasal cavity. This tissue, called the olfactory epithelium, contains millions of smell receptor cells all housed in a remarkably tiny area of about one square centimeter per nasal cavity. The epithelium contains four major types of cells: ciliated olfactory receptor cells, microvillar cells, supporting cells, and basal cells. The cells are nourished by the lamina propria, the layer of connective tissue that binds the epithelium to the bone or cartilage underneath.

Each receptor cell housed in the olfactory epithelium is adorned with microscopic hairlike projections called cilia. The cilia are like tentacles floating in a layer of mucus. They are also greedy for scent molecules. Proteins on the cilia reach out in attempts to latch onto passing odor molecules. Once the odor molecule and the receptor cell hook up, the reaction is buzzed to the olfactory bulb at the base of the brain, just above the roof of the nasal cavity. From there, the signal travels to the brain and the odor is identified. Though seemingly complex, an odorant is really just a molecule that possesses a particular shape. When we inhale a molecule, it searches for and binds to the appropriate receptor sites.

Damage to the olfactory epithelium can cause temporary or permanent loss of the sense of smell. According to the Olfactory Research Fund, *anosmia* is a complete absence of the sense of smell; *hyposmia* is a diminished sense of smell; *dysosmia/*

parosmia is a distortion of the sense of smell; *phantosmia* refers to detecting the odor of something that is not there; and *heterosmia* causes all odors to smell the same.

A person experiencing problems with his or her sense of smell may suffer from a number of disorders—neurological (Alzheimer's and Parkinson's disease), hormonal, or viral. Sometimes, upper respiratory infections can interfere with the ability to smell. However, head injuries account for the most serious cases of olfactory impairment. Approximately 20 to 30 percent of head trauma victims lose some or all of their sense of smell. A severe blow to the back of the head can jar the brain inside the skull, shearing the axons of the ciliated receptor cells. The olfactory epithelium becomes disorganized and is no longer able to send its signals. Fortunately, the neural tissues of the olfactory epithelium (as well as the epithelium of the VNO) can regenerate themselves and, in some cases, reinstate the sense of smell.

The Olfactory Research Fund suggests that when people meet, they engage in "olfactory bonding." Just as we carry around and disseminate our pheromones, we also have a "smell fingerprint," a unique body odor that is the cumulation of the food we eat, our individual genetic codes, the makeup of our skin, the medications we take, the kind of mood we're in, and even the weather.

Smells, like pheromones, help us to define our models of attraction. Who could argue with French poet and novelist Rémy de Gourmont, who said: "The woman one loves always smells good."

The five senses are what allow us to make sense, literally, of our surroundings. Now that we have a knowledge of the physiological origin and workings of the sixth sense, we have the opportunity to understand ourselves and other people more fully. Why are we motivated to do certain things? Why do we react to situations and people in sometimes unpredictable ways? The answers to such questions may be revealed in the delicate and ever-complicated brain dance of the senses.

Sex, Love, and Lust: The Pheromone Connection

This fatal attraction, impersonally called chemistry and attached to subliminal pheromones, has its autonomy of force apart from both genetics and environment.

—James Hillman, *The Soul's Code*

Walking in a Cloud

Pheromones are our constant companions. We are surrounded by a shadow self because of the chemically volatile nature of our pheromone molecules, which allows them to travel freely through the air, and because we shed thousands of pheromone-containing skin cells every day.

When our pheromone cloud bumps up against another person's pheromone cloud, one of two things can happen: Our clouds want to mingle and form a bigger, even better, cloud, or they want to float away into their respective skies.

Think about a passionate encounter or relationship you once had or are now in. Write down your first impressions of the situation and describe, if you can, your physiological (body) reactions. Did the meeting take place at a party, over dinner, at the office watercooler? How did the events unfold? How did you feel around the other person? Did you feel instantly connected to the person standing in front of you? Did your heart start to beat quickly? Did sweat drip from your palms? Maybe you blushed persistently or couldn't get to sleep that night.

As you were pulled by the magnet of sexual chemistry, your vomeronasal organ and hypothalamus were in the director's chair orchestrating a complicated production staged inside your body. A whiff of the person's pheromones snapped your VNO into action. The organ sent the pheromone signal promptly to your hypothalamus, eliciting some kind of response: *Stay near this person or, better yet, get closer!* or, *Fall in love (or lust) with this person!* or, *Turn and run. Right now! End this conversation about what's on the best-seller list and walk to the other side of the room!*

Andreas Capellanus wrote in his twelfth-century treatise *The Art of Courtly Love* that love "gets its name (amor) from the word for hook (amus), which means 'to capture' or 'be captured,' for he who is in love is captured in the chains of desire and wishes to capture someone else with his hook."

As the hooks of attraction, pheromones can be subtle at times, outspoken at others. An instant reaction to someone may be powerful enough to attach you for life in an erotic bond cemented by good chemistry. Or your encounter might take longer to come to a boil, starting with a slow simmer and building with applied heat. Pheromones can also clash like the opposing ends of two magnets: No matter what you do, say, or think, the connection refuses to be made.

The Case for Smelly T-Shirts

In a laboratory in Switzerland, a woman pokes her nose into a box. She sniffs, then sniffs again. She moves to box after box, inhaling. Is she testing the latest brand of perfume destined for the cosmetics counter? In a way, she is. The woman is breathing in the thick aroma of men's sweaty, unwashed T-shirts saturated with odor (and odorless pheromones) released from the men's underarm apocrine glands.

It's all part of an experiment conducted by zoologist Claus Wedekind, Ph.D., of the University of Bern in Switzerland, to monitor the reactions of forty-nine female volunteers to garments worn by forty-four males. Before handing his cotton T-shirt over to science, each man had slept in it for two consecutive nights and had avoided deodorants, colognes, scented soaps and lotions, spicy foods (which can affect the aroma of sweat), alcohol, and sex.

The women selected for the experiment were all in the middle of their menstrual cycles, that time of the month when a woman's sense of smell is particularly keen—some say one hundred times as sharp as normal. And, for two weeks before the experiment, the women's noses were treated with a special nasal spray designed to coat and protect the fragile mucous membranes of the nasal walls, so that their sniffing abilities would not be inhibited by any external factors or injuries.

To test his theories, Wedekind placed each sweaty shirt inside a box with a hole cut in the side and asked the women to indicate their preferences by scoring each shirt based on the categories of "intensity of smell," "pleasantness," and "sexiness."

The results caused a stir. The experiments revealed that the women were the *most attracted* to the odors of shirts that had been worn by men whose immune system genes were the *most different* from their own genetic makeups. What this means is that when a woman goes to choose a mate, she subconsciously does so by "sniffing" out his immune system. Thus, she will prefer the mate whose immune system is the most unlike hers (his odor will be the most attractive to her).

The Darwinian logic behind the women's T-shirt preferences is that when two *dissimilar* sets of genes are combined, the result is a more comprehensive immune system for the pair's offspring. Outcrossing breeds progeny that are less prone to illness because they possess more varied and robust immune system genes and thus exhibit "hybrid vigor." While the conventional wisdom is that a woman will select a mate based on his external qualities—personality, education, background, ambition—

Wedekind's study shows us that she will be more likely to choose her mate based on how appealing she finds the scent of his perspiration.

How does the T-shirt study relate to pheromones? It reveals the intimate link between pheromones and mate selection because of this fact: *Immune system genes are closely allied with pheromone-producing genes and are located on the same chromosome.* The connection between immune system genes and a person's distinct body odor and pheromonal makeup was first addressed by biologist Lewis Thomas in 1974. An award-winning science writer whose works include *The Lives of a Cell* and *Late Night Thoughts on Listening to Mahler's Ninth Symphony*, Thomas theorized in an essay titled "Fear of Pheromones" that each person's one-of-a-kind pheromonal signature is also a "printout" of that individual's immune system genetic code. Scientists call it the major histocompatibility complex, or MHC. MHC genes, which number about fifty, are situated on a single chromosome and are, according to the Monell Chemical Senses Center in Philadelphia, "the most variable in all of nature—they differ greatly from individual to individual."

This phenomenon of females' attraction to mates with dissimilar immune system genes wasn't new to the world of science, as it had already been proven to occur in laboratory mice. But the discovery that human females also favor mates with "opposite" genes shed new light on how pheromones can affect our reproductive objectives. When a woman favors one unwashed T-shirt over another, she sends a clear message about her own genes and what she requires in a potential father of her children. She also broadcasts her likes and dislikes about the man's pheromones and body odor. An odor or pheromone perceived as negative is almost guaranteed to turn a woman off.

The extraordinary thing about MHC readings is that they are not processed consciously. A woman who finds one pheromone-laden shirt offensive and another pleasant or sexy makes her choice without relying on her thinking brain for guidance or advice. The women in Wedekind's experiment couldn't say *why* they preferred certain sweaty shirts (and genetic codes) over others, only that they had definite attractions to some and strong repulsions to others.

As he was reviewing the results of the T-shirt study, Wedekind made yet another noteworthy discovery. Some of the women were not drawn to the odors of their genetically optimal matches, but instead to odors of men with genes similar to their own. These women had one thing in common: They were all taking birth control pills.

This is a fascinating, and even disturbing, thought. The synthetic hormones in the Pill trick the female body into thinking it is pregnant by preventing ovulation. A "pregnant" woman is not likely to be out searching for a mate, but she may be drawn instinctively to men with *similar* genetic codes. Why? Wedekind thinks it may be because men with similar genes resemble the woman's family, her protectors, her clan—those who can support her while she carries her child to term.

Is it possible that women who take birth control pills are not capable of choosing their best mates? Doctors already know that couples who lose children to miscarriages often have similar MHC readings instead of the genetically preferable dissimilar MHC lineup. And the children of MHC-similar couples sometimes have lower-than-average birth weights. Conceiving a child can also be more difficult and take longer for MHC-similar couples.

Can dissimilar MHCs and their inherent pheromonal components truly bond a couple? Pheromones are the instigators of romance and love, but they will not guarantee a lifelong union because successful relationships rely on the interplay of many factors.

Cuddle up to His Armpit

Forget his chiseled chin, well-defined arm muscles, and impossibly green eyes. The sexiest part of a man may very well be his armpits.

With their abundance of pheromone-producing apocrine

glands, the armpits play a key role in sex, love, and lust. The link between the apocrines, human pheromone communication, and human sexuality has caught the attention of scientists investigating how pheromones might be able to affect our physiological processes.

One scientist with a fascination for the human armpit is Dr. Winnifred Cutler. Cutler earned her doctorate in biology from the University of Pennsylvania and did postdoctoral work in behavioral endocrinology at Stanford University. She also founded and operates the Athena Institute for Women's Wellness, Inc., in Chester Springs, Pennsylvania. Cutler writes in her book *Love Cycles* that frequent sexual contact with a man "seems to promote fertile-type reproductive cycles in women." It took some time for Cutler to determine how this might be possible, but she eventually arrived at a theory: male pheromones.

Cutler's idea that male pheromones might affect the menstrual cycle (and therefore lead to fertile reproductive cycles) was already known to occur in animals. In a study conducted by Dr. Martha McClintock (whose discovery of menstrual synchrony among human females we'll see in chapter 6), it was shown that the body odors and pheromones of male rats could enhance the fertility of female rats in the vicinity.

Working with organic chemist Dr. George Preti at the Monell Chemical Senses Center, Cutler designed an experiment to find out if so-called male "essences" (secretions from the armpit that contain pheromones and sweat) would have an effect on female endocrine (hormonal) and reproductive cycles.

To summarize the experiment, Cutler gathered armpit perspiration from male volunteers who wore cotton collection pads under their arms. She extracted what she thought were the key ingredients from the secretions and froze these essences. She then selected female participants for the study, choosing only women whose menstrual cycles had been historically irregular; that is, they did not follow the typical pattern of onset approximately every twenty-nine days, plus or minus about three days. Also, each woman recruited for the experiment had not been engaging in frequent (several times a week) sex with a man.

In the first stage of the experiment, Cutler swabbed the upper lips of the female volunteers with the thawed male essences. She repeated the application several times a week for about fourteen weeks. The results of the study seem to suggest that frequent contact with a man—even if this contact is in the form of his underarm sweat worn on the lip—can help to put a woman's irregular menstrual cycle on track. Indeed, most of the participating women who had had irregular or aberrant-length menstrual cycles and who did not have frequent sexual relations with men began to have regular-length menstrual periods after three-and-a-half months of lip swabbing. Even though these women had not been sleeping with men, they began to display the fertility attributes of a sexually active female. A woman's reproductive ability is based in part on the regularity and length of her periods. Cutler concluded, "The male essence seemed to substitute for regular weekly sex."

Maybe this is good news for women whose mates travel frequently for business or for those women or men who just aren't interested in committing to sex several times a week. Based on Cutler's findings, it appears that all a woman has to do is collect her husband's or boyfriend's underarm perspiration, dab it under her nose (which sends the pheromone signal to her VNO and from there to her hypothalamus), and enjoy regular menstrual periods and possibly enhanced fertility.

Cutler reminds us that a woman can take full advantage of the benefits of male pheromones only if she is in *close physical contact* with a man. In other words, merely hanging out with men is not enough; there must be some intense "nose-to-armpit" contact to allow the man's pheromones to reach the female's pheromone receptors in her VNO. Therefore, a woman should kiss a man, make love with him, inhale the scent of his sweat, or sleep with her head near his armpit to get the full power of his pheromones.

If a heterosexual woman unknowingly relies on the pheromone-perfumed sweat of a man to help keep her menstrual cycle on track, what happens with lesbian couples? Based on research conducted by the Kinsey Institute and her own studies, Cutler found that women who are sexually intimate with other women also experience more evenly regulated menstrual cycles. However, there is one major difference: A homosexual woman requires *three times* as much sexual activity as a heterosexual woman does to achieve the same result in her menstrual cycle. Cutler writes, "Sexual relations three times a week among lesbian women were equivalent to one time a week among heterosexual women."

"An Intolerable Neural Itch . . ."

This is how the poet W. H. Auden described the feeling of sexual craving. But craving, as we are learning, may be nothing more than a chemical reaction compelling us to seek sexual intimacy and closeness.

Remember chemistry class and the hours spent poring over glass beakers, adding, subtracting, mixing, and swirling? If we paid attention and maneuvered through the experiment properly, our reward was chemical activity. The contents of the beaker would begin to smoke or change color like a liquid chameleon and we would readjust our safety goggles, a bit nervous in the presence of such stormy molecular movement.

Whether you adored or abhorred chemistry is not the issue. The point is this: Chemical reactions occur in our bodies every day, all day. When we experience such a reaction, we're often shaken by the sheer intensity of it. In the movie *Ninotchka*, Greta Garbo said, "Love is a romantic designation for a most ordinary biological—or, shall we say, chemical?—process." Some of the strongest reactions are facilitated by the powerful emotions and the associated physiological effects of love and lust.

Talk to anyone who has experienced intense love or lust and he or she will recount a similar experience. When suddenly in love, people often find themselves emerging from a mental or physical lull. Others work themselves into such an emotional roar that they have trouble sleeping, focusing on their work, or being in any situation or place from which the beloved is absent.

People in this state are high, their brains awash in love chemicals let loose by the neurochemical floodgates.

But humans are not automatons. We don't respond to just any stimulus. Unlike the many animals and insects who react to pheromonal messages reflexively, we usually don't find ourselves unexplainably in the throes of passion at the instant we meet someone with irresistible pheromones. Still, there are times when another person's chemistry threatens to overwhelm our logic. This is what it means to be "blind with ecstasy." The thinking brain stops "seeing" and the love-hungry hypothalamus takes over.

New research into why we are attracted to some people and not others points to the phenomenon of chemistry. We say, *Greg and I had instant chemistry—there were fireworks from the beginning* or, *When I met Marsha, I wanted to be with her every chance I could get. She was like a drug*. We may also recall finding someone physically pleasing to the eye and socially desirable (secure job, nice home, well dressed, financially stable) but noticeably lacking in the attraction department—thus, the fireworks are absent and the flame languishes, unstoked.

This may seem confusing. If someone possesses the characteristics society deems worthy, wouldn't those attributes suffice in our courting and mating rituals? The answer is no. We would be more accurate, and infinitely more human, to describe our partners this way: "Our chemistry was on target."

In the 1970s, sex experts Masters and Johnson told us that sex followed a predictable pattern of four stages: excitement, plateau, orgasm, and resolution. The news of the four-cycle sex tango became material for lively dinner conversation everywhere, but something was missing from the picture: What causes or at least facilitates sexual excitement and interest?

Another sex expert, Helen Singer Kaplan, M.D., Ph.D., seemed to provide that answer when she announced that sex incorporated a fifth element: the spark, or, more accurately, sexual chemistry. Without this basic initial stage of connection, one might find it difficult to progress through the subsequent four levels. In her 1974 book *The New Sex Therapy*, Kaplan did mention pheromones as possible instigators of sexual attraction, but she called them odors and connected them to the sense of smell. She also made what is now thought to be an inaccurate correlation between pheromones and aphrodisiacs; pheromones are not aphrodisiacs but rather subtle communicators that provide information and not instant, flat-out lust. Nevertheless, Kaplan laid the foundation for future discussions of pheromones as they might apply to human sexuality when she wrote, "[There is] no doubt that a tantalizing aroma is a powerful aphrodisiac, even if not consciously recognized."

We now know that sexual attraction and mate selection have a physiological connection to pheromones. Pheromones are the flames of desire. They can also be the fire extinguishers of dislike or repulsion. If two people find that their pheromones are in harmony and transmitting messages of mutual attraction and desire, then sex is more likely to occur.

Pheromonal cues are exchanged under our noses. We aren't consciously aware of what's happening when pheromones zap our brains with pulses of desire. Unlike the more above-board workings of our other senses, pheromone perception doesn't announce its existence. You can't smell a pheromone.

Regardless of how subliminal pheromones are, there is no mistaking their power. Would you fall in love with someone with whom you had no chemistry? A successful and enduring match between two people requires a host of extrinsic factors, but at the deepest level it relies on the sixth sense.

Looking Back on Love

People have been intrigued by love for centuries. That intrigue has supplied writers and poets through the ages with plenty of material. Poet Rainer Maria Rilke was frank and to the point when he wrote about love in *Letters to a Young Poet*: "For one human being to love another is perhaps the most difficult task of all, the epitome, the ultimate test."

We usually prefer to think of our relationships with intimate others as romantic rather than purely chemical, and this is a preference that has been handed down over the years. Most of us would rather focus on the higher attributes of romance and courting than the basic human drives that spring from chemical impulses.

What images come to you when you hear the word *romance*? High-minded ideas about romance began to form when the troubadours, poets and knights living in Europe during the twelfth century, entered the scene. The troubadours were members of the nobility. They had the luxury of time to craft their poems and music. They weren't as interested in sexual intercourse as they were in the art of courtly love. In fact, it didn't matter if a relationship stopped just short of consummation, because sex wasn't the goal of this coy game.

Looking back on the lives of these creative lovers, we could say that the troubadours were successful at switching their pheromones to the "off" position. This is perhaps because the troubadors were most interested in what Joseph Campbell calls the "psychology of love."

The notion of romantic love has survived into the modern age. The first American Valentine's Day card was distributed on February 14, 1667, adorned with a picture of a rose. Today, many of us still equate romance with red roses, red wine, swelling music, walks in the moonlight. Anthropologists suggest that red is the assigned color of romance and love because it makes a subconscious reference to the genitals, which redden as they engorge with blood during sexual activity. Red is also the color of a blush, a sign, says zoologist Desmond Morris, of nubility and sexual readiness. That could explain why American women spend some $875 million on cosmetic blush every year.

On the flip side of romance, a number of cultures around the world have traditionally complicated and sometimes sneaky methods of "inducing" love in the opposite sex, usually with human body fluids such as semen, vaginal secretions, saliva, perspiration, and urine, all of which are thought to contain pheromones. Women in Victorian England could earn money by peddling handkerchiefs perfumed with their own body odors.

The only rule to this form of entrepreneurism was that the woman herself have a pleasant smell; otherwise, sales would understandably lag. According to one old recipe, a man would do well to mix some of his semen with sugar and then, while his love interest isn't looking, add the substance to her drink. Legend has it that once the potion touches the lady's tongue, she will swoop upon the nearest man in unleashed passion. If you are a woman, you are advised to wipe the end of your love interest's penis with an article of your clothing. (How you might accomplish this without him noticing is not explained.) Regardless of your method, your goal is to collect a sample of the man's semen during the wiping process. Bury the specimen beneath your door. Your man will then come running to you, professing eternal love—or so the story goes.

On a more up-to-date note, if you were to conduct an Internet search of the word *pheromone*, you might land at a site dedicated to the practice of incorporating human urine into the sex act. Enthusiasts claim urine is an underrated aphrodisiac, and that because animals often spray each other with urine when aroused, humans can do the same to intensify their lovemaking.

Pheromones' Partners in Sex, Love, and Lust

We know that pheromones have powerful effects on the bodies of men and women. A woman's reproductive cycle even takes cues from the pheromones and odors found in a man's armpit, and pheromones can help us to choose our optimal mates. But pheromones themselves do not make up the entire picture. When we meet someone and fall in love or lust, our bodies go into overdrive as they react to signals sent by the brain.

We produce a number of other hormones and neurochemicals when we are struck by Cupid's arrow. Hormones are produced by the endocrine glands (for example, the pituitary, adrenals, testes, ovaries, thyroid, and pancreas) and then released into the bloodstream. Once there, hormones deliver their instructions to specific target cells, where the message is converted into the appropriate response. Neurochemicals originate in the brain and spend their time traveling the delicate pathways of the synapses. Here are some of the more familiar substances that are featured in the sexual repertoire of the human body.

Phenylethylamine (PEA)

When we fall in love, our brains begin to manufacture increased quantities of phenylethylamine (PEA). This amphetaminelike molecule is a feel-good endorphin that makes our brains buzz and increases our blood pressure and heart rate. Thus, the "high" we feel when in the beginning stages of infatuation, that sense of steady euphoria, is rooted in the biology of the body, specifically in PEA. It is a true human love drug. When we fall in love or lust, our blood levels of PEA surge.

A discussion of PEA brings us invariably to the subject of chocolate. The confection's popularity has led researchers to investigate why we crave it and buy it by the ton. The eighteenth-century Swedish biologist-taxonomist Carolus Linnaeus named the cacao tree *Theobroma*—the "food of the gods." Sixteenth-century Spanish explorer Hernando Cortés brought chocolate to the "civilized" world. Cortés ultimately overthrew the Aztec rulers of Mexico, but before things got ugly, the Aztecs introduced the Spaniard to a hot drink called chocolatl. That was in 1519. Like many of us, Cortés fell in love with chocolate.

One recent survey revealed that 70 percent of female respondents would rather have chocolate than sex. According to the Chocolate Manufacturers of America, each American consumes an average of eleven pounds of chocolate every year. That seems like a lot of candy, until you consider the sweet-tooth habits of the Swiss, who each consume an average of twenty-two pounds of chocolate per year!

Why the chocolate obsession? It's simple, really. Cocoa, a key ingredient of chocolate, contains PEA. Chocolate is also thought to boost levels of depression-alleviating serotonin and other endorphins in the brain. Some people become addicted to sex, others to chocolate.

Fortunately, serious chocolate addicts (those who binge and purge repeatedly) can get help from an opiate-blocking drug called naloxone. The drug acts by preventing chocolate-inspired endorphins from completing their work. Adam Drewnowski, Ph.D., director of the Human Nutrition Program at the University of Michigan, treated women with naloxone and documented a dramatic reduction in bingeing behavior. Naloxone gets between the endorphins and their receptors in the brain, breaking the feel-good link.

Oxytocin

Oxytocin is the "cuddle chemical," the molecule that inspires a reaction when we touch or are touched, however briefly, by someone we love. A short polypeptide hormone, oxytocin is released by the pituitary gland. In addition to its touch-related actions, oxytocin stimulates the contraction of the smooth muscle of the uterus during labor and orgasm and facilitates milk flow during nursing.

Without oxytocin, our intense feelings of attachment to our children or to our sexual partners or spouses would dissolve. Our skin, which relies partly on oxytocin for its sensitivity to touch, would feel dead without this chemical. When a mother hears her hungry baby cry, oxytocin intervenes, making her nipples erect and ready to pass the stream of milk so desired by the infant. A touch from a loved one causes oxytocin to be released into the bloodstream. A glance at the object of our affections can also result in an oxytocin surge. Sexual intercourse is guaranteed to send the body's oxytocin production rates into warp speed.

Vasopressin

Sex therapist and researcher Theresa Crenshaw, M.D., calls vasopressin the "monogamy molecule." Indeed, when it works in conjunction with the fiery male sex hormone testosterone, vasopressin has a tempering effect that keeps males' sex drives in check and stops them from following their hormones straight to the next attractive female. It puts the "sense" back into the man.

Like oxytocin, vasopressin is secreted by the pituitary gland. It constricts blood vessels, raises blood pressure, concentrates urine by moving water back into the bloodstream, and is present in large quantities during REM sleep. Also called *antidiuretic hormone,* vasopressin aids in regulating our body temperatures and helps men and women rein in their tempers.

Testosterone

Testosterone is the quintessential male hormone. It makes men *men*. It gives them square jaws, hairy chests, deep voices, and muscular bodies. Women also produce testosterone, but in much lower concentrations.

In males, testosterone is produced primarily in the testes and the adrenals and it facilitates the development and maintenance of secondary sex characteristics. In women, it is produced in the ovaries and the adrenals and it sensitizes the genitals and nipples to pleasurable touch. It also plays a part in a woman's sex drive. Testosterone production tapers off during menopause, leaving some women with deflated libidos. One treatment uses supplemental testosterone to boost the flagging sex drives of menopausal women. The average woman has some 40 nanograms (one billionth of a gram) of the hormone in each deciliter of her blood. Men, by contrast, have anywhere from 300 to 1,000 nanograms per deciliter.

A recent study by Dr. Christina Wang of the University of California at Los Angeles found that testosterone may not deserve its reputation as the male aggression hormone. Rather, men *deficient* in testosterone display the negative attitudes (irritability, testy mood) normally associated with the presence of the hormone. Before being treated with testosterone supplements, the fifty-four hypogonadal (low-testosterone) men in Wang's study expressed a range of negative behavior, including being overly aggressive, a trait historically associated with an *abundance* of testosterone. Testosterone replacement therapy improved the moods of the men in Wang's study. They became friendlier and calmer, not exactly the demeanor one would expect from an increase of the hormone.

Testosterone plays an important role in sexual behavior, particularly in a man. It is the hormone that jump-starts his libido, giving him the drive to pursue sexual intercourse, and is also involved in the production of sperm. A vegetarian diet is thought to reduce testosterone levels, and a losing streak in a man's life can also cause a drop (getting fired, his wife leaving him for another man, financial hardship).

Even watching a favorite sports team lose a game can lower testosterone levels. Research conducted by Georgia State University psychology professor James Dabbs, Ph.D., found that after a sporting event, male fans whose team won had surging levels of testosterone, while the fans of the losing team had decreased testosterone levels. Dabbs monitored the reactions of twelve Brazilian and nine Italian male soccer fans seated at a bar watching a final play-off between their respective countries. The Brazilian team won; as a result, the South American fans' testosterone levels increased by 28 percent. The losing Italians, however, showed a 27 percent drop in their testosterone levels.

Estrogen

Estrogen is the classic female hormone. Produced chiefly by the ovaries, estrogen is responsible for the development of female secondary sex characteristics. It gives a woman her sensuous curves, softening the angles of her body with the folds of femininity.

Estrogen affects body fat stores, salt and fluid retention, and blood clotting, to name a few of its functions. For many years its primary work was thought to involve the regulation of the menstrual cycle, but it is now known that estrogen also plays a part in the maintenance of bone density, mental functioning, and the health of the heart. In the sexual arena, estrogen helps make a woman receptive to sexual advances. It is what gives a woman her own personal body scent, and it is crucial in vaginal lubrication and the texture of the skin.

Endocrinologist Bruce S. McEwen, Ph.D., of Rockefeller University found that rats exposed to supplemental estrogen displayed more activity in the neural synapses between the brain's hippocampus and the hypothalamus, the part of the brain that regulates sexual behavior. The theory is that estrogen not only makes a woman feel and smell female but is capable of promoting sexual interest and pursuit as well.

Dopamine

Dopamine affects how we see and react to life. Do we experience pleasure in an embrace with a lover, in a salmon-colored sunset, or in the sugary mist of newly fallen snow? If the answer is yes, then dopamine is at work. If things of beauty and love leave us feeling as flat as old soda, then something's wrong with our dopamine levels. Dopamine facilitates our addictions, whether to alcohol or great sex, because it sends us out in search of pleasure. Dopamine is an unabashedly hedonistic chemical. It gives us the intense pleasure of orgasms. No wonder we can't resist its seduction.

Dopamine is a neurotransmitter produced in the brain. Because of its strong effect on the sex drives of both genders, it is an important chemical in the sex, love, and lust connection.

Serotonin

Serotonin, a neurotransmitter formed from the amino acid tryptophan, is found in the brain, blood serum, and gastric mucous membranes. It is active in vasoconstriction, stimulation of the smooth muscles, transmission of impulses between nerve cells, and regulation of cyclic body processes. When serotonin levels are high, sex drive is dampened. When levels are low, sex drive is heightened.

Our bodies are like factories; they are always at work producing the many chemicals that make us who we are and influence the nature of our relationships. Now, we will switch gears for a moment and discuss another very important non-chemical component of attraction.

The Look of Love

When we separate love and lust from their inherent pheromonal connection, we find other ways of viewing attraction. We might deny being drawn to someone based purely on how he or she looks, but the truth is that certain facial and body traits are more appealing to us than others.

A relatively new science has evolved to determine why some people are considered attractive and others are not. It involves measuring body proportions, in particular the symmetry of one's features, to pinpoint the parameters of desirability.

Newsweek recently featured a cover story on physical attractiveness titled "The Biology of Beauty: What Science Has Discovered About Sex Appeal." The article cites the work of University of New Mexico ecologist Randy Thornhill and psychologist Steven Gangestad and includes a photograph of popular actor Denzel Washington. The theory is that Washington is universally considered attractive because his face lines up; in other words, you could draw imaginary horizontal lines between various points on his face (cheekbones, jawbone, lips, nostrils, outer eyes) and the vertical line created by the midpoints of the horizontal lines would form an even stretch. By contrast, singer Lyle Lovett's mug is a jumble. His face is asymmetrical, and his overall physical attractiveness rating is lower as a result.

University of Texas psychologist Devendra Singh notes that while a person's face is a reliable indicator of whether he or she will be considered attractive, body shape and proportion also deliver powerful messages that trigger desire and longing.

Singh focused on what men find attractive in women. When shown a series of drawings depicting women of varying physical proportions, from underweight to normal weight to overweight, men find themselves drawn to a certain body type. In this study, women with waist-to-hip ratios of 0.6 to 0.8 (their waists are 60 to 80 percent the size of their hips) were the most attractive to men. This ratio doesn't depend on a woman's weight; if she has the desirable ratio, then she will most likely be considered attractive. The classic female figure—36-25-36—has a waist-to-hip ratio of 0.7.

Why would a man be drawn to a woman who possesses the physical measurements that form the desirable hourglass shape? A female with a slim waist and full hips is visibly capable of producing children, so the thinking goes. And this thinking has merit. A 1993 Dutch study of women seeking the services of a fertility clinic found that the women's waist-to-hip ratios affected their chances of conceiving. The women whose waist-to-hip ratios were between 0.7 and 0.8 were more successful in their attempts to get pregnant.

Randy Thornhill and Steven Gangestad also found that a woman's ability to achieve orgasm is related to the symmetry of her mate's features, and the most desirable men—and those with whom women reach orgasm more quickly—are those with the most "even" features.

Symmetry is such an important component of attraction that each year almost half a million Americans (including some fifty thousand men) employ the services of plastic surgeons. The goal is to improve the balance and symmetry of their facial features.

Humans aren't alone in their preference for beautiful partners. Animals, too, size up the looks of their potential mates. While pheromones play key roles in the sensory lives of animals and insects, symmetry is also an integral part of the mating game. For example, a scorpion fly with symmetrical wings will more easily attract mates and have better luck securing food than will his lopsided peer.

D. H. Lawrence wrote, "Sex and beauty are inseparable, like life and consciousness." In the human quest for love and lust, it appears that beauty lies not only in the eye of the beholder, but in the genetic program that influences perception of beauty as a factor in reproductive success.

Kiss Me

Who could deny the pleasure of a well-executed kiss? Soft meets soft, the personal space bubble opens its doors, the guard is let down. Kissing is about as close as you can get to another human being without removing your clothes.

The romantic tingle of a kiss is undeniable. It conveniently steps in and fills the space of unspeakable passion, as Ingrid Bergman told us when she said, "A kiss is a lovely trick designed by nature to stop speech when words become superfluous." The Romans thought so highly of kisses that they had several words in their Latin lexicon to describe the touching of lips: an *osculum* is a friendly kiss applied to the cheeks, a *basium* is a kiss planted with a bit more affection, and a *suavium* is an ardent kiss reserved for lovers.

Some theories say kissing evolved out of our ancient need to "sniff out" strangers, much the way animals do. (Is this why Eskimos rub noses in greeting?) Animals spend the majority of their time sniffing around; in the process they receive massive doses of olfactory and pheromonal stimulation.

Back when humans consciously sniffed each other (we still do it, but not as overtly), we also engaged in a subtle exchange of breath. This exhaled air is sometimes called the *mana*, an invisible representation of a person's deepest self. Edwin Dobb writes in *Harper's* magazine, "Kissing was considered the only unmediated way to mingle souls, and no manner of lovemaking could be more intimate or more consequential."

Mingling of souls aside, kissing provided our ancestors with clues about other people—what they ate, the state of their health, their personal hygiene or lack of it. In the quest for procreation, kissing was, and still is, a way to rule out undesirable mates. You wouldn't make babies with someone whose breath was foul with disease, would you?

Researchers have found that couples who kiss frequently have stronger relationships and more satisfying sex lives than their nonkissing peers. Kissing boosts feelings of security and well-being, and displaying and receiving affection can strengthen the immune system. In 1992 Janice Kiecolt-Glaser, Ph.D., of the Ohio State University College of Medicine, compared the immune function of ninety newlywed couples. By collecting samples of the volunteers' blood, she could determine how affectionate the couples were toward each other. Those who kissed and cuddled on a consistent basis had higher levels of antibodies and virus-killing T-cells coursing through their bloodstreams than did the "cold" couples.

Sex educators Drs. Bob and Leah Schwartz advise couples to "use all five senses—and even your brain—when you kiss." But a kiss becomes more complete and ultimately more satisfying when we also take into account the subtle presence of the sixth sense. In the language of tribes residing in Burma and Borneo, "kiss" means "smell," an allusion, perhaps, to the ability of the kiss to deliver messages to the deepest regions of the brain.

When viewed in the context of pheromones, the simple kiss takes on new significance. This is because the narrow strip of skin between the top lip and the nose, the nasal sulcus, is a pheromone powerhouse—it produces a vast quantity of chemical messengers. The pheromone-receiving VNO lies just inside the nose. When you kiss someone, your nose nuzzles his or her upper lip, which allows your VNO to receive the pheromones housed there. And the person kissing you is simultaneously receiving your pheromones. Thus, a kiss is most certainly not just a kiss; it is a way to get closer to another person's pheromones and to inhale those molecules deep into the nose—and into the realm of the sixth sense. Now that we know pheromones are so highly involved in the kiss, it makes even more sense that this act of love and affection is still a favored aspect of the human connection, particularly the connection between lovers.

Shakespeare called sex "a madness, most discreet, a choking gall, and a preserved sweet."

Who hasn't experienced soul-stirring sexual chemistry or the tapping of feisty neurochemicals deep inside the brain? Being in love is like drinking a miraculous elixir, one that makes you feel younger, more alive, more desirable and attractive.

Your pheromones, in turn, debrief you about your potential love interests. They kindly let you know if something wonderful stands to happen or if the relationship is a dud destined for a distant corner of the memory rather than the bedroom.

More Pheromone Mysteries

A man that smells good . . .

—the singer Madonna, on what she finds attractive

Your husband still makes your heart pound after a decade of marriage. Is it his handsome face, his way with children, his steel blue eyes? Or, is it his pheromones?

You are madly in love with a woman for the first time in your life. Is it her smile, her way with words, the way she dresses? Maybe her pheromones draw you to her.

You are so in tune with your newborn baby that you find yourself stopping several times a day to contemplate the bond between you and your child. Is this connection facilitated by the intimacy of the birth process, or by the fact that you are the child's primary caregiver? Or, is it that you and your infant share an unparalleled ability to communicate with pheromones?

You can't stand to be in the same room with the neighbor across the street. As you wrestle for a reasonable explanation for this ongoing dislike, the intuitive side of your psyche reminds you that the situation may just be the result of bad chemistry. You don't like this person's pheromones.

Why do we fall for a particular person, or form lasting friendships with certain people and despise others? Why does a mother instinctively know her baby? These are not easy questions to answer, and new questions surface as we journey farther into the world of pheromonal communication and the sixth sense.

The mysteries of human attraction and interaction continue to confound us. Sometimes we struggle to assign words and thoughts to our experiences. Ongoing research into human behavior has given us some answers, but despite this progress we still search for insights. We crave information about ourselves, about why we do what we do.

But pheromones can be a bit frustrating at times. Part of this could stem from the fact that pheromones are odorless. Unlike odors, which we process consciously with our sense of smell, pheromones, which we process unconsciously, are not easy to define. They can't be labeled, because we can't explain with precision what pheromones mean to us and do to us. But when we step back from our need to define and categorize events, situations, and experiences, it dawns on us that the elusive nature of pheromones doesn't detract from their significance. Even if we can't pin down the workings of pheromones with absolute clarity, we don't deny their presence. Who would be bold enough to discount the intuition of the sixth sense? Would you dispute that you have feelings and thoughts that seem to be governed by a sensory system quite different from your senses of smell, taste, touch, hearing, and sight? Would you refute the presence of chemical signals, or pheromones, in your life? Would you insist that you've never felt good or bad chemistry? The true sensory nature of the sixth sense has only recently been discovered and tested. But you've lived with your sixth sense for all of your waking and sleeping hours, even before your mother gave birth to you.

Of course, to say that pheromones are at the forefront of each and every human interaction and function would be simplistic and scientifically inaccurate. The aim of this book is to open your mind to *possibility*, to the idea that pheromones can affect you on levels that go far beyond the realm of anything science can at this time illuminate with certainty.

Different cultures around the world give us colorful examples of how pheromones might influence human behavior and interaction. Also, common human experiences, such as dancing, may have much to tell us about pheromone communication. Here are some interesting points to ponder:

- During active royal festivities and events, France's King Henry III is said to have wiped his face dry with a sweaty garment belonging to his beloved Marie of Cleves.
- In New Guinea, some tribal members say good-bye to each other by inserting a hand into the other person's armpit and smearing the gathered sweat over their own bodies.
- People of Asiatic descent have fewer underarm apocrine glands than do Occidentals. As a result of this, Asians have a lighter body odor than people of European descent, and they often find Westerners' odors too strong and offensive to the nose.
- In Nigeria and among African-Americans in the southern United States, some women have been known to feed their unfaithful husbands or lovers a special soup that contains their menstrual blood and vaginal secretions. The thinking is that if a woman's lover has strayed, whether in reality or in his mind, he can be lured back to her by ingesting her bodily fluids.
- In the Caribbean, a person attempting to attract a sweetheart will feed him or her a hamburger patty containing an unusual ingredient: sweat from the cook's armpits. After ingesting the delicacy, the person on the receiving end will supposedly fall in love with the creative chef.

- An unmarried man living alone on a remote island notices that when he visits the mainland and spends time with women, his facial hair grows faster.
- When people dance, they also perspire, which allows their pheromones to be released into the air. Sometimes, a fast dance is followed by a slow one during which the woman moves in toward the man's body, her face and nose close to his armpits.
- Blonds, brunettes, and redheads are said to possess distinctly different body odors, and men often express preferences for the color of hair they like on a woman. Women, too, have their own ideas about what is attractive in a man. Could these preferences be related to classes of pheromones genetically linked to hair color?
- The benefit of oral-genital sex has intrigued anthropologists and evolutionary scientists for years. But if we look closer at the physiology of oral sex, we find that it might have significance in mate selection.

A preference of hair color, an attraction to sweat, dancing, oral sex, body odor—all are examples of how pheromones might be able to influence our interactions with other people. Is a preference for a specific hair color based on the pheromones and body odor of the person we admire? When we say we love dark hair, are we really saying we like people whose body chemistries produce pheromones that are most pleasing to our vomeronasal organs?

Strands of hair make an especially practical way for pheromones to deliver their messages. Pheromones cling to strands of hair, only you can't see or feel them. That's why our underarms and groin areas (and, in men, the chest, which can be hairy) are so heavily imbued with pheromones; we hold them close in our axillary and pubic hair. When the skin is warm and moist, as it is in the groin and armpit areas, pheromone-containing sweat from the apocrines becomes highly volatile, and therefore more of our pheromones can ostensibly travel into the air and into the vomeronasal organs of the people we encounter. But while the hair on our heads might not be as pheromone-concentrated as the hair under our clothes, the fact that our hair is out in the world means our pheromones are out there, too.

Each one of us has a unique body chemistry and a one-of-a-kind pheromonal fingerprint. But is it possible that people with light hair and complexions have a body odor that sets them apart from those with dark hair, who have their own signature body scents? Men with dark hair sometimes have darker skin than their lighter-haired peers. While Asians have fewer apocrines than whites, blacks have more than both. While whites usually don't have the glands on their chests and abdomens, blacks often have a number of them in these regions. People with dark hair and skin naturally have more apocrine glands throughout their bodies, so it may be that the allure of dark hair has something to do with increased pheromone production.

We know that pheromones are players in the universal game of love and lust. We also know that pheromones are not the singular reason why we fall in love or don't. When you begin to feel an attraction to someone, you use a number of criteria to select a mate. One is to look at the person's physical makeup, or symmetry (see chapter 5). When we say we like blond hair, we may be responding to sixth sense pheromones, but we could also be accessing another resource: What sex researcher John Money calls *lovemaps*. The director of the Johns Hopkins University School of Medicine's psychohormonal research division, Money based his lovemap concept on studies of human attraction conducted in a variety of cultures. According to Money, lovemaps develop from our childhood experiences, whether positive or negative. If as children we developed a fondness for men who wear eyeglasses, we are likely to carry that preference into adulthood and into the realm of our mate selections.

Money believes we draw our lovemaps early; for most of us, the maps are drawn by the time we celebrate our eighth birthdays. Our lovemaps are subconscious representations of our experiences with family members, friends, our parents' acquaintances, teachers, and so on.

How trustworthy are lovemaps? Do they guarantee eternal bliss? We can't live by pheromonal chemistry alone, which is why we might fall madly in love with a sexy musician whose pheromones make us drool, but we won't necessarily marry him and have his children. Lovemaps do not take into account the important role of chemistry, and therefore pheromones, in our relationships. What if we followed our lovemaps to a fault, only to find ourselves repeatedly in relationships that lack spark and passion? We can't deny the pull of chemistry, as writer Nan Goldin reminds us in her essay "The Ballad of Sexual Dependency": "Even if relationships are destructive, people cling together. It's a biochemical reaction, it stimulates that part of your brain that is only satisfied by love, heroin, or chocolate; love can be an addiction." It seems clear that as we search for the perfect partner, many factors must intertwine in the tapestry of love and romance.

The love-as-addiction definition shakes our foundation and sends tiny cracks through our carefully laid groundwork, because we know that the addictive properties of love don't always bode well for long-term partnerships.

Hidden Pheromones

Oral sex is perhaps the most intimate encounter you can have with another person. It is a way of saying, "You're okay. You're clean and healthy and I want to be close to you." But, oral sex has not always been widely condoned. In the *Kama Sutra,* the ancient Hindu text on sexuality, love, and marriage, oral sex is described as being practiced by "unchaste and wanton women, female attendants and serving maids."

The Romans, in spite of their hedonistic ways, frowned on oral sex unless it was performed between homosexual men or women. Heterosexual men were allowed to receive oral sex from the prostitutes they hired, but, as Diane Ackerman writes in *A Natural History of Love,* "it was considered repulsive and degrading for a man to pleasure a woman with his mouth." Ackerman further explains that Roman men did not put their mouths on women because they had concerns about sanitation; rather, they refrained from performing oral sex on a woman because the act was thought to compromise their masculinity. Giving rather than receiving oral sex made a man appear submissive. It made the woman the dominant participant in the sex act.

Nonetheless, the practice of oral sex has survived, and some theories, particularly those that have evolved in the field of psychology, hold that oral sex is a way for grown-ups to satisfy the oral fetishes that began to take shape in childhood and which supposedly influence adult behavior. Sigmund Freud defined oral sex (along with kissing and fondling) as a practice that adults embrace in their attempts to reconnect with the pleasures they received while breast-feeding or being cuddled at their mothers' bosoms. Freud also commented on the key role of genital odors in the sexual relations of men and women, noting that some people believe these odors to be sexual stimulants. In Freud's world, however, "civilized" people deemed such smells offensive and openly expressed their disbelief that anyone would find these scents even marginally appealing.

Freud and his colleague Wilhelm Fliess studied the relationship between the nose (they referred to the sense of smell and not to the pheromone sense) and the sexual organs. Physiological research has shown that when a woman is having sexual intercourse, when she is menstruating, or when she is pregnant, her nasal environment becomes more sensitive and more prone to nosebleeds. Fliess called such alterations "genital localizations." His theories about smell and sexuality are discussed in his book *The Relationship Between the Nose and the Genital Organs of the Woman*, which was first published in 1897.

Since the pubic hair traps pheromones, we subscribe to the theory that humans engage in oral sex out of a subconscious desire to inhale a lover's pheromones. Pheromones are highly concentrated in the genital regions, and the only way to get the

VNO close to these molecules is to place the nose right where they live.

A study conducted at Emory University provides insight into how vaginal secretions might affect sexual response. Researchers there looked at how male rhesus monkeys respond to aliphatic acids, the vaginal secretions of female monkeys. These secretions are also called copulins because they induce copulatory behavior in males. A female rhesus secretes the greatest quantities of copulins when she is at her most fertile stage; that is, when she is in her reproductive prime and, like human females, considered the most desirable to the opposite sex.

Some scientists believe aliphatic acids in vaginal secretions may function as pheromones in primates and humans, although this theory has not yet been confirmed. When a male rhesus monkey sniffs a female monkey's copulins, he becomes aroused and eager to have sexual intercourse with her. The male rhesus will display this behavior even when the vaginal secretions are from a *human* female. If a human female's aliphatic acids are applied to the vaginal area of a spayed female rhesus, the male will try to initiate intercourse with the animal.

Historically, vaginal fluids have been incorporated into courting rituals. Egyptian women and women living in eighteenth-century France placed a linen bag in their vulvas that would, in time, become perfumed with their vaginal secretions; they would then use this bag to attract the attentions of men. French prostitutes applied vaginal fluids behind their ears to lure male passersby into their beds.

Today's women's magazines and how-to sex manuals advise lovers to place an emphasis on hygiene, however, and they suggest bathing or showering before making love. Such cleansing rituals are meant to keep sexual intercourse as odor-free as possible, almost as if making love to someone who actually smells like a human and not a floral display is repulsive. But does bathing deprive our VNOs of the pheromones that are thought to be produced in the most intimate regions of our bodies? Because we have been conditioned to find some odors pleasant and others offensive, we may be failing to connect with a crucial link to our humanity. Our pheromones exist for a good reason—to help us communicate with other people. We might be better off not focusing so much on removing our pheromones as on learning to accept them, and their accompanying odors, as necessary attributes of who we are.

Why Hair Is Sexy

Many American men find a sleek underarm on a woman sexy and alluring, and some are turned off by unshaven armpits. Has the standard practice of shaving armpits—and thus removing pheromones—turned American women into shadows of their true pheromonal selves? If such men could rework their definitions of what is attractive and what is not, would they come to accept, even prefer, hairy women? Would contact with women's concentrated pheromones have a positive effect on men? Do men in European countries, where shaving underarm and leg hair is not as rigidly practiced, enjoy more pheromonal conversations with their mates?

And what about facial hair on men? There are women who would sooner smooch with their pet dogs than kiss a man with a full beard and mustache. But, this doesn't stop men from growing their facial hair into amazing configurations. According to a survey conducted by the Wahl Shaving Research Center and Remington Products, 56 percent of American men have facial hair, but 51 percent of women say they are "repulsed" by hair on a man's face. While 46 percent of men say a mustache or beard gives them an aura of maturity and 32 percent say facial hair makes them appear more distinguished looking, only 11 percent say facial hair turns up the wattage on their sex appeal.

The nasal sulcus, the strip of skin between the top lip and the base of the nostrils, produces large quantities of pheromones thanks to its abundance of sweat glands. Would coaxing their stubble into mustaches or full beards make men more manly, sexier, more pheromonal? When a woman kisses a hairy-faced man, does she take more of his "trapped" pheromones into her VNO?

The Mystery of Menstrual Synchrony

If you are a woman who has spent time in close quarters with other women, you may have experienced a strange phenomenon. Whether the setting was a college dormitory, a sorority house, or a busy office, you may have noticed that, after a few months, you and your friends began to get your menstrual periods at the same time.

What would cause a group of women living or working together to fall into synchronized menstrual rhythms? The answer lies in the remarkable powers of the sixth sense. Some research suggests that odorless pheromones, floating through the air in invisible waves, can affect and alter the onset of a woman's menstrual cycle.

The mystery of menstrual synchrony has long been discussed by women who experience it, but it wasn't addressed comprehensively in the scientific world until 1971. Martha McClintock, an undergraduate psychology student at Wellesley College, set out to investigate why women who spend time together often menstruate together. McClintock, living in an all-female dorm at the time, observed that she and her close friends slowly came around to getting their periods at the same time. Even more fascinating was that the women's menstrual cycles tended to lock onto the cycle of a "dominant" woman. McClintock's mind began to churn: Was this synchrony a fluke? Or was it connected to the fact that the women lived together? Were their individual body chemistries responsible for the shift? Was it possible that one woman, through her own dominant chemical signals, could unknowingly force other women to menstruate with her?

To find out, McClintock monitored the menstrual cycles of 135 female volunteers. She found that in the space of just a few months some of the women fell into menstrual synchrony while others did not. McClintock noted that the women who experienced synchrony were more than just living in a dormitory together; they were roommates and close friends who spent a good deal of time in each other's company.

McClintock then collected sweat samples from a "regulating" woman's underarms and dabbed the substances on the upper lips of female volunteers. Women in a control group received applications of alcohol to their upper lips. After four months of lip swabbing, the women's periods had fallen into sync with the woman who had the dominant body chemistry, while the control group maintained their normal cycles.

McClintock's findings were published in the scientific journal *Nature*, where she concluded that a woman's ability to bring menstrual cycles into sync with her own might be facilitated by pheromonal communication—or, in this case, manipulation. Her finding is referred to as the McClintock Effect.

Menstrual synchrony has been the subject of numerous studies since McClintock published her findings nearly three decades ago. In 1993 scientists in the Department of Psychology at Israel's Bar-Ilan University documented menstrual synchrony between mothers and daughters, women sharing a room in a private home, and women living in close quarters in a dormitory. Of particular interest is the scientists' finding that the highest degree of menstrual synchrony was between mothers and daughters.

Pheromone researchers Winnifred Cutler and George Preti, whose "male essence" study is discussed in chapter 5, also found that the menstrual cycles of women whose upper lips are swabbed with armpit secretions from other women can fall into sync with the cycle of the pheromone "donor." The scientists exposed ten women with normal menstrual cycles to the underarm sweat of female volunteers. After some three months of regular lip swabbing, menstrual synchrony occurred between the volunteers and the donors.

The phenomenon of menstrual synchrony shows how pheromones can exert their effect on the hypothalamus, the region of the brain that controls the body's hormonal system and therefore menstruation. Hormones regulate how long a woman's period will last and when it will return, and the hypothalamus oversees this process. The menstrual synchrony studies seem to indicate that the pheromones of one "dominant" woman can affect the hormonal regulation of other women.

Pheromones also appear to play a significant role in delaying the start of menstruation. One study found that girls attending same-sex boarding schools had later onsets of puberty than did their peers who went to coed schools and thus had more contact with the opposite sex. The reason for this is still unknown, but it may be linked to male-female contact and the effect of such proximity on the reproductive system.

Social Butterflies?

Pheromones influence how we feel by sending signals to the "emotional" hypothalamus that help us react to people and situations. They can even affect our moods. A study designed by researchers at the University College of London's psychology department and the developmental neurobiology unit at London's Institute of Neurology showed how pheromones may be able to elicit changes in human behavior. The study was led by John Cowley at Hatfield Polytechnic in Hertfordshire, England.

The researchers enlisted thirty-eight female volunteers and thirty-eight male volunteers. All were asked to wear a necklace overnight, keeping it on while they slept and not removing it for any length of time until the next morning. They were told the necklaces, made from lightweight plastic tubing, contained "harmless substances." The volunteers were not given any information about the nature of the study or the mysterious substances housed in the necklaces. What the volunteers didn't know was that their necklaces contained a cotton swab soaked in either vaginal secretions or human androstenol, a putative pheromone present in underarm sweat. The volunteers in the control group wore necklaces with swabs soaked in chloroform.

The next morning, the volunteers were asked to evaluate their moods and, specifically, their interactions with other people that had occurred while they were wearing the necklaces.

They were asked to recall each person they had spoken to that morning, in the hours between waking up and filling out the questionnaire. Any interaction was included in the results. The volunteers carefully documented all of their verbal exchanges with other people, noting how long each interaction had lasted, under what circumstances it had occurred, and who had initiated the conversation.

The male volunteers whose necklaces contained either androstenol or vaginal secretions did not report any significant differences in their interactions with men or women, and the female volunteers who had been exposed to androstenol or vaginal secretions did not note any drastic alteration in their interactions with other women.

But the experiment did reveal an effect on the behavior of the female volunteers who had unknowingly spent seventeen hours in the company of odorless androstenol vapors: They logged significantly higher numbers of social exchanges *with men*. The women said they felt more comfortable approaching men and that they had initiated conversations more frequently than they might have done under normal conditions.

In their written summary of the experiment published in the *Journal of Steroid Biochemistry and Molecular Biology*, the researchers explain, "In general terms, the study has clearly demonstrated the capacity of a substance with known pheromonal properties, 5α-16-androsten-3α-ol, to affect the social behaviour in a natural setting of young women exposed to its vapour overnight. The women exposed to the 16-androstene pheromone exercise (not necessarily consciously) showed an increased responsiveness to men." The study suggests that "exposure to androstenol altered the women's behaviour in such a way as to intensify exchanges with men. The pheromone might perhaps operate by making the women feel more at ease with men, and/or more attractive to them or more attracted by them."

Another experiment, also led by Cowley, investigated how pheromones might play a role in shaping our impressions of other people. In this study, three men and three women competing for student office were brought into a room, and after each candidate made a presentation, volunteers were asked to rate the candidates on their perceived abilities to be effective leaders. Each volunteer wore a surgical mask; they were told the purpose of the mask was to disguise their faces, so that their facial expressions would not be visible to the political hopefuls. In reality, the masks of half the volunteers had been imbued either with androstenone, a molecular variation of the pheromone androstenol, or human vaginal secretions.

The male volunteers wearing the treated masks responded to the candidates in much the same way as did the male volunteers wearing the untreated masks. The female volunteers whose masks contained vaginal secretions preferred candidates who seemed shy and modest, and they gave the aggressive candidates low marks. But the female volunteers wearing androstenone-treated masks preferred the outgoing, bold candidates over the quiet, unassuming ones and awarded them the top marks.

These studies are interesting and provocative because they suggest our moods and behaviors can be directly affected by pheromones. Pheromone researcher David Berliner suspects that some people may even be able to detect another person's moods by "reading" his or her airborne pheromonal cues. He hasn't tested his theory yet, but he has spoken to several people who have unusually sensitive VNOs. These individuals have told Berliner that they can "translate" a person's mood, but don't know how they can do this. They are sure only that they are powerfully affected by the presence of other people. Berliner thinks this ability may have more to do with pheromonal communication and the sixth sense than extrasensory perception.

Without relying on other sensory cues—is the person frowning? Is his voice strained? Is he perspiring more than normal? Is he fidgeting?—people with highly tuned VNOs may be able to read a person's mood based solely on the pheromones he or she is sending into the air.

Show Me Some Skin

Those of you who live in places where winter stretches across the majority of the calendar know firsthand the glory of replacing wool and down with T-shirts and shorts. Wonderful things can happen when we uncover our skin. One study found that summertime sex is the steamiest, the most erotic, and occurs more frequently than in the quieter months of winter. We sweat profusely in the summer, too, and this sweat does more than cool our bodies; it also holds our pheromones and allows them to waft into the air and then enter the VNOs of our friends, acquaintances, and the people we pass on the street or stand next to in line for the movie. We're not as grumpy in the summer, either, because extended hours of sunlight trigger bursts of mood-elevating estrogen in both men and women.

Research into how a person's address can affect his or her personality and moods reveals that people living in the southern regions of the United States and in similar latitudes in other countries are often friendlier and more open than northerners, who are characteristically cooler, less ebullient. One study in particular illustrates the behavioral characteristics that occur across cultures and in countries that have markedly different weather patterns. It was conducted by psychologist James Pennebaker and his colleagues at Southern Methodist University, in conjunction with Bernard Rime of the Catholic University of Louvain in Belgium.

The researchers interviewed three thousand college students in twenty-six countries, asking them to describe their own levels of emotional expressiveness and what they perceived to be the expressiveness—or taciturnity—of people living in other regions of the countries they called home. In eighteen out of twenty-six countries, students' responses meshed with the personality traits that seem to define certain geographic regions: Students living in warmer climates described themselves as more outgoing, friendly, cheerful, and social than did the students from more frigid northern regions. Such differences surfaced among students living in countries as diverse as Japan and Italy. The researchers also found that if a person relocates, his or her attitude and demeanor will adapt to meet the social environment of the new home. This means that a happy-go-lucky southerner might experience a change in personality with a move to the north, and vice versa.

In searching for a reason behind climate-controlled personality differences, Pennebaker points to the fact that northerners wear more clothing and are less able to display body language or read the body language of their friends and neighbors, which may lead to hindered social skills. He also posits that because people living in cold places are forced to stay indoors more often than their warm-weather peers, they naturally become less social.

But if we look at these differences in the context of pheromones, we arrive at an intriguing thought: Could it be that the friendly attitude of southerners has something to do with their pheromones, which are free to float around in the warm air, unrestricted by heavy winter clothing? Are northerners cooler because they aren't getting enough of their fellow humans' pheromones?

Pheromone researcher Louis Monti-Bloch shared with us his theory about that perennial human affliction, spring fever: "We release pheromones constantly. In the spring and summer, people wear less clothing as the weather warms up, therefore more skin is exposed and people are receiving chemical inputs they are not aware of." The way Monti-Bloch describes it, spring fever could herald the arrival of a pheromone army into the air we breathe, which could explain why many of us drift around in a state of bliss as the months advance toward summer. When we cover our skin with clothing, we block the pheromones secreted through the skin from circulating through the air. This fact seems to validate our desire—however embarrassed we might be by its acknowledgment—to show some skin. Go ahead and wear that slightly risqué dress or tank top. Men and women are designed to communicate with one another in ways that go beyond the trappings of words and other sensory signals. We can speak volumes through the lovely, complicated canvas of our skin. But we do have to make an effort to engage in the conversation if we are to reap the rewards.

The Mother-Baby Bond

Picture a fetus suspended in its mother's womb. It wants for nothing. Its environment is perfectly climate-controlled, its appetite is sated at a moment's notice, and its daily rhythms are determined by an ancient code designed to keep it in a happy state of equilibrium.

It is always a treat to watch a mother-to-be communicate with her developing baby. She pats and caresses her stomach, plays music to the small being in her body, coos and sings softly. After the baby is born, people will marvel at the degree to which the mother and infant have already cemented their bond. Onlookers will notice that the infant often squawks and screams until it is in the arms of its mother; no other form of comforting will assuage the baby.

A mother has the uncanny ability to recognize her infant by how he smells and, some scientists think, by his pheromones as well. And a newborn can identify the smell of his own mother even if he can't see her or hear her voice. Infants being prepped for surgery are often swaddled with things that smell like their mothers—a blouse, a blanket, a pillow case. As the baby dozes under the anesthetic, it feels protected and nurtured in the presence of its mother's scent.

What gives a baby the ability to identify its mother in a room full of people, even when the child is just a few days old? What compels a baby to crave its mother's—and only its mother's—breast? In his writings, Charles Darwin noted that an infant will instinctively turn its head toward its mother's nipple; he attributed the behavior to a recognition of either the mother's body odor or the comforting heat of her body. Pheromone researcher William C. Agosta cites a study in which researchers blindfolded new mothers and brought in three infants: the mother's own and two others. Agosta writes, "In 61 percent of the trials, the mother correctly identified her own baby, much better than the 33 percent expected by chance."

The pheromonal link between mothers and babies has been documented in animals, too. Several studies conducted by animal behaviorist Frank Beach showed that mother rats can identify their own babies even if their offspring are disguised in another litter. However, pheromones can have negative effects on the maternal behavior of mammals. A non-pregnant female rat with "dominant" pheromones will at times storm around the cage like a hellion, eating the babies of the other females in the pack and stealing their nests. Researchers at the University of Chicago found that this destructive behavior can be subdued and even stopped if the ill-tempered rat gets to know the pregnant rats before the babies are born. Howard Moltz and Julie Mennella have described a special pheromone that a pregnant rat releases which, when perceived by the other female, makes her less aggressive and squelches her desire to kill and eat babies.

One theory of prenatal communication between the offspring and its mother holds that while in the womb, the fetus receives olfactory information via the amniotic fluid; it is thought that this form of communication is what facilitates the baby's recognition of its mother by her smell after birth. If the fetus receives olfactory cues from its mother, it is likely that it also receives pheromones from her, too. When the baby is born, he would ostensibly know his mother because he had been reading her olfactory messages and getting to know her pheromone chemistry for months. The baby can't be fooled, either. In one experiment, six-week-old infants were exposed to the armpit pheromones of female volunteers. When the babies were in the presence of the pheromones of women other than their mothers, they either cried or ignored the stimulus.

When a baby nurses at its mother's breast, he is exposed to the pheromones that emanate from her nipple, areola, and armpits. It has been found that if a mother washes one breast with soap and water before nursing and leaves the other one unwashed, the baby will seek the unwashed breast, for it carries the comforting pheromones and odor of his caretaker.

This discussion brings us to the breast-feed versus bottle-feed debate. Some women cannot breast-feed because they experience intense pain or do not produce enough milk. Others work full-time and don't have the luxury of nursing their babies all day. Still others don't want to breast-feed their infants in public. Does a baby fed from a bottle suffer from no contact with its mother's breast pheromones, and would this have an impact on its future development? Is the mother-baby bond weakened if the baby never gets to put its mouth around its mother's nipple?

A New Look at Tribal Bonding

Mother and child may communicate via pheromones. Women who sleep next to men reap reproductive benefits just by inhaling their lover's pheromones. People who live together get steady infusions of each other's pheromones. Pheromonal communication remains an important part of our lives even as we go to great pains to stake out our individuality and guard our personal space.

The American Dream was born out of this need to be separate, and it created a people who prefer detached homes over apartment buildings and ten acres of their own land rather than a common garden. Even people who seem content living in cities get dreamy looks in their eyes when they fantasize about a home in the middle of nowhere, with no distractions, human or otherwise.

But it's important to remember that despite our notions of civilized society and our ever-growing storehouse of technological developments, we are still tribal creatures who need contact with each other. Somewhere in our brains we remember the days when communal living kept us in close quarters.

"We are tribal primates," says David Moran. "As a society, we've forgotten that fact. But, I think people are now starting to remember the tribal aspects of themselves and the role of pheromones in tribal living."

Some of us live alone, work alone, travel alone, eat alone, and sleep alone. But in the deepest corners of our brains, where ancient memories have been stored for thousands of years, we crave closeness. When we can no longer stand to be alone in the world, with only our own pheromones for company, we crowd ourselves into restaurants, dance halls, stadiums, churches, theaters, shopping malls, and amusement parks.

In recent years there has been a resurgence of group activities: "Sensitive New-Age Men" meet to male-bond and swap stories; book lovers meet to discuss what they've read (while alone, most likely); and self-help and encounter groups encourage much hugging and physical contact.

That we still seek the company of other people, even though we could by all accounts sustain ourselves independently, speaks to the true nature of who we are. In this age of intense individualism, we still yearn to be close to *something*. So, we leave our isolation behind for a few hours of human contact. We eat, we shop, we are entertained, and we engage in more than a few chemical conversations along the way.

Some scientists studying how humans communicate via pheromones believe tribal bonding has a strong pheromonal component. In countries where children don't automatically get their own bedrooms, families live right under each other's noses. They eat, sleep, read, and talk in the same room day after day, year after year. Pheromones are everywhere, and they probably contribute to feelings of attachment and love.

While the pheromonal link to tribal bonding has not been conclusively established by scientific studies, the theory makes sense. Pheromones are a form of subtle chemical communication, a way to "read" another person or group of people, and the closer we are to people, the more tuned in to them we can be. Says ethnomedicine scholar Terence McKenna in the book *Trialogues at the Edge of the West*, "I think that pheromones are vastly underrated for their organizing power in biology and social systems."

Pheromones and Homosexuality

The controversial debate about what causes homosexuality in humans continues. Some claim homosexuality is genetically determined, while others believe a persuasion for members of one's own sex is more affected by how a person is raised and what he or she was exposed to as a child. As the fetal brain develops, numerous hormones come into play that determine the baby's gender. These hormones are also thought to affect the brain wiring of the fetus. If the hormones are thrown off kilter, the baby's sexual orientation can be altered. But if homosexuality is biologically determined, could pheromones also be involved?

British neuroanatomist Simon LeVay has written extensively about the possible origins of homosexuality. His research has led him to his theory that homosexuality is determined by genetic factors. Specifically, he studied the hypothalamus, homing in on the INAH-3 nucleus, which he thinks could be linked to a sexual preference for women. He found that this nucleus varies in size depending on a person's sexual orientation. The nucleus is significantly smaller in homosexual males than it is in heterosexual males, and more similar in size to the nucleus of a heterosexual female.

One study looked at how the chemical composition of urine might indicate who is heterosexual and who is homosexual. Researchers collected urine from male volunteers and tested the samples for levels of androsterone and *etiocholanolone,* both of which may have pheromonal properties. The findings were surprising: The urine of the male volunteers who divulged later that they were homosexual contained higher concentrations of etiocholanolone and lower concentrations of androsterone than did the urine of the heterosexual males. In the heterosexual men, just the opposite occurred: There was more androsterone than etiocholanolone present in their urine. Subsequent studies have supported these results. Androsterone and etiocholanolone are byproducts of testosterone and other androgen breakdown by the liver. Low levels of androsterone and other androgens reveal again and again a preference among males for members of their own sex, lending credence to the theory that sexual preference is determined in the womb.

(An interesting sidebar to this finding is that heterosexual women and homosexual men have similar concentrations of etiocholanolone and androsterone in their urine. This points to the possibility that a male homosexual's production of androgens [the hormones that create male sex characteristics] is more like a female's than a heterosexual male's; this would explain why he would not exhibit "normal" male behavior.)

David Moran recounts an experience he had while working as a scientific consultant for Human Pheromone Sciences, a company that manufactures Realm-perfumes containing synthesized human pheromones. The women's fragrance contains more male pheromones than female pheromones, and the male version contains more female pheromones than male. The formulations are designed to give the wearer a dose of the pheromones of the opposite sex. On several occasions, homosexual men told Moran they preferred the women's version of the perfume over the male version.

While the fact that some homosexual men express a preference for a perfume containing a high percentage of male pheromone is anecdotal, it points to the possibility that homosexuality may be linked to pheromone production and perception.

Moran notes that a male preference for a male pheromone could involve differences in the VNO, the hypothalamus, and their interconnections. Perhaps future studies into the link between sexual orientation and the sixth sense will give us answers.

Fear and Intuition: Pheromones, Too?

A new field of study has caught the attention of the media: how people can tap into their intuition to protect themselves from becoming crime victims.

Called "violence prediction," this new science focuses on teaching people how to identify someone who may be a potential assailant. Every situation is different, but experts point to similarities that can act as guidelines in assessing potentially dangerous scenarios: body language, voice inflection, speech patterns, and how language is used to intimidate and create a false sense of intimacy. The basic theory underlying violence prediction is that we know on a subconscious level if someone is not quite right. Without realizing we're doing so, we pick up on subtle cues in a person's speech and body language.

Gavin de Becker, a violence-prediction expert whose book *The Gift of Fear: Survival Signals That Protect Us from Violence*, has captured the attention of readers intent on finding out how they can avoid becoming targets of crime, contends that fear itself is a good measure of potential problems. De Becker claims we are hard-wired to process the subtleties of fear in amazingly accurate ways. He describes it this way in an interview in *Redbook*: "You feel fear. The doors to the elevator open and you see a man inside who causes you to feel apprehension. Maybe it's the way he looks at you, his size, his eyes—it doesn't matter why. But your intuition speaks to you in clear terms, telling you there is danger. . . . Your intuitive capabilities have an extraordinary capacity to protect you. . . . Intuition is the cornerstone of your safety. It's nature's ultra-efficient way of alerting you, instantaneously, to danger."

Once again we hear that word: *intuition*. In order to ward off danger, we're told to listen to our intuition, to follow our gut feelings. We are well-advised to heed our silent but important warnings about other people. When someone frightens us, we are encouraged to embrace that fear rather than rationalize and dispose of it. But there is also a physiological component to fear.

Our bodies undergo profound changes when we are frightened, angry, or stressed. Dogs can "smell" fear because of their ability to detect very slight changes in a person's body chemistry. People who are gearing up to commit a crime experience increases in body temperature, heart rate, and blood pressure. They also emit a strong, musky odor, the by-product of a nervous, amped-up sweat produced by the apocrine glands.

Given that the body reacts physiologically to the prospect of commiting crime, could it be that our intuition and gut feelings are really responses to pheromones? When we meet a stranger in an elevator and immediately get the willies, are we experiencing a negative reaction to this person's body chemistry? Perhaps a person about to commit a serious crime like rape, murder, or robbery emits a pheromonal signal that reflects his internal state. The body, revved up to do harm to another person, could actually change its own chemistry, and the victim might quite literally be able to sniff such a warning.

The theory of violence prediction involves using all of one's sensory systems to assess someone who is, for whatever reason, a bit off. We might look at this person and notice a shift in his eyes as he speaks, or our sense of hearing might discover inconsistencies and alert us accordingly.

But when we are face-to-face with a potential attacker, is it our sixth sense that truly sends the warning message? When we feel nervous around someone, uneasy for no reason and tempted to flee rather than flirt, are we being driven by pheromonal signals relayed to our hypothalamus, which controls our emotions and our autonomic fight-or-flight response?

Science has not yet given us definitive answers to these questions, but it is intriguing to consider the possibilities. Yes, we should, as de Becker advises, attempt to be as keyed in as possible to the people we encounter. A part of that awareness means being highly attuned to our sixth sense.

Life Without a VNO

What would happen if you couldn't sense another person's pheromones? Life would lose a good portion of its mystique and color. Without the ability to detect pheromones, it's possible that humans would lack the anatomical hardware to make healthy decisions about their lives and loves. Perhaps you would fall in love with the wrong person over and over again, or find yourself in situations that could have been avoided with input from your sixth sense. Without your sixth sense and your ability to decipher other people's pheromones, you wouldn't be operating on all of your sensory cylinders.

People born with atrophic gonads and non-functioning olfactory systems suffer from a congenital condition called Kallmann's syndrome. Louis Monti-Bloch and colleagues found recently that patients who have this condition are also born without a vomeronasal organ.

In a groundbreaking study that served to further highlight the importance of the human VNO's role in sexual development, Monti-Bloch, David Berliner, and fellow scientist Vicente Diaz-Sanchez studied ten male patients with Kallmann's syndrome and found that four of them indeed *did* have a sense of smell. Why is this significant? For many years, Kallmann's was thought to be the result of olfactory impairment alone. But, this study illuminated a startling fact: A percentage of Kallmann's patients may actually have a working olfactory system, but in *all cases* they lack a functioning VNO. The researchers, writing in their scientific abstract, "Absence of Vomeronasal Organ (VNO) Function in Patients with Hypogonadotropic Hypogonadism", explain, "Our results show that in all ten . . . patients, the vomeronasal-terminalis was non-functional. However, four of these patients had a functional olfactory system, which does not conform with the definition of Kallmann's syndrome present in the other six patients. . . . We are describing for the first time a syndrome characterized by hypogonadotropic hypogonadism with a functional olfactory system and without a functional vomeronasal system."

Kallmann's syndrome is thought to occur when the hypothalamus fails to make adequate quantities of GnRH—gonadotropin-releasing hormone. GnRH regulates the release of sex hormones in the pituitary gland, adrenals, testes, and ovaries. Without the correct migration of GnRH neurons to the brain during the fetal stage, the hypothalamus is unable to send the appropriate signals to the endocrine system, and sexual underdevelopment results because the necessary amounts of sex hormones have not been created.

A person born with Kallmann's (and without a VNO) will reach puberty later than his normal peers. When his friends begin to show an interest in the opposite sex, he will feel unusually flat in the attraction department. While the hormones of his classmates are taking off like rockets, his will sit quiet. If he begins to date at all or show an interest in sex, it will happen much later than would be expected. Kallmann's syndrome affects the individual for life: it influences whether he will fall in love or lust, even whether he will marry and have children. Men with Kallmann's have smaller-than-average genitals and lack normal facial hair growth. Kallmann's syndrome is not as common in women as it is in men, but when it does occur, it results in sparse pubic hair and underdeveloped breasts. Women with Kallmann's do not menstruate.

In chapter 3 we learned that the human VNO is housed close to, but is separate from, the olfactory system. However, these two sensory routes work in parallel through their common conduit, the nose. Like the sense of smell, which allows us to discern the myriad odors we encounter every day, the VNO is a crucial piece of our anatomy because it helps us navigate through a world teeming with pheromones. If we had the misfortune to have been born without a VNO, we would be unable to experience the world fully.

Studies involving male and female rodents have shown that removing the VNO results in drastic hormonal suppression and permanently altered or abandoned mating patterns. Because the neural tissues of the VNO can regenerate, complete removal of the tissue is necessary to accurately document what happens when the VNO is removed. Leaving even a tiny portion of VNO epithelium (tissue) intact can fool the pheromone-receiving system into thinking the organ is still functional.

Pheromone researchers Charles Wysocki and John Lepri

found that removing the VNO of a male mouse left him unable to respond as he would normally to the sexual cues of a female of his species. A male mouse with an intact VNO usually experiences a boost in luteinizing hormone (LH), which stimulates testosterone production when he is exposed to the urine of the female mouse. This initial surge in LH causes the male's testosterone levels to spike, giving him the desire to engage in sexual activity. If he doesn't have a VNO with which to process the female's chemical sex signals, or pheromones, his reproductive behavior and ability to procreate are either greatly diminished or halted. Another study found that male guinea pigs whose VNOs had been removed stopped exhibiting a common behavior: wagging and bobbing their heads in response to the pheromone-laden urine of the female.

At the time of this writing, no experiments had been conducted to determine what would happen to a person if his or her VNO were purposefully removed. This is not surprising—who would volunteer for such a study? Nevertheless, surgeons who repair nasal deformities or conduct rhinoplasty should be careful to avoid the accidental removal of the patient's VNO during the surgery.

Don't Cut That Nose!

In 1991 plastic surgeon Dr. José Garcia-Velasco, working with his colleague Manuel Mondragon in the department of reconstructive surgery at the University of Mexico School of Medicine in Mexico City, looked for the VNO in 1,000 rhinoplasty patients (579 women and 421 men). He found what he was looking for by probing gently with a nasal speculum inserted into the patients' noses and peering inside their nasal cavities with the assistance of a headlamp or, if needed, a special fiber-optic device that added magnification.

Garcia-Velasco located a symmetrical pair of VNOs in all but 192 of his patients. Of those 192, 125 had septal pathologies—twisted and distorted nasal septums that obscured the organ. He corrected the septal deviations and looked again. This time, he could clearly see the VNO in 102 of the 125 people who had had the corrective surgery. Garcia-Velasco and Mondragon also discovered that the VNO is similar in appearance and frequency in men and women.

Impressed with their findings, they began to worry that patients undergoing nasal surgery might be losing their VNOs simply because their surgeons are not aware of either the organ's existence or its importance. They wrote: "These findings, together with data from other electronmicroscopic and neurophysiological studies, should be taken into consideration because of possible problems that might result by performing surgical alterations on the nose without preserving the VNO.

"Thus, we need to start thinking about preserving the VNO in any nasal operation, especially in the surgical correction of the twisted nose associated with severe septal deviation."

Every year, hundreds of thousands of people undergo nasal surgery, either to correct a health problem or to improve their appearance. The hope is that Garcia-Velasco's warning will be heeded by plastic surgeons and that many VNOs will be saved as a result.

More Food for Thought

We know that pheromones help us to make decisions about the people in our lives. However subconsciously we may process those chemical messages, they are important to us nonetheless. But, what happens to us when we go about our business without the one-on-one communication that facilitates the workings of the sixth sense? Here are a few scenarios for you to think about:

Many Americans work from their homes. Are people who work alone, away from their "tribe," subject to pheromone deprivation? Would they be better off working in offices, where pheromones float up and down the corridors and into conference rooms? Is our fascination with the Internet and communicating via computerized chats and e-mail putting us at risk of extreme sensory deprivation, especially where the sixth sense is concerned? Is our technology-obsessed culture isolating us from our own humanity? Fax machines allow us to send and receive information without leaving our offices or homes. We can rent movies instead of going to theaters packed with people. And, of course, telephones give us instant communication capabilities but do not allow us to use the priceless gift of being able to talk with our sixth sense.

Most of us own automobiles and shy away from using public transportation. If we continually encapsulate ourselves in the confines of our own spaces, how can we get to know our fellow humans through the sixth sense–activating power of pheromones? Does taking a bus or train to work give us something important that we aren't even aware of?

Distance learning, in which students receive instruction via videotapes, television, or the Internet, could have some real drawbacks. When students of any age sit together in a classroom, they are surrounded by their own pheromones and those of their classmates. Is there a negative impact to learning at home in relative isolation because of the lack of chemical communication with other people?

Pheromones seduce us even as they elude us. We know they're there, but we don't fully understand their effects on our minds and bodies.

There is something comforting about discovery and knowing that leads us to new ways of thinking about what it means to be human. But if we step back for a moment, perhaps trim our mental sails and chart a new course, we find that there is also something undeniably pleasing and intriguing about the unknown.

We tend to put more weight on facts and numbers than on theories and hypotheses, but our human nature tells us that some of what makes us who we are can't be explained. As pheromones exert their subtle influences on our lives, they teach us something about ourselves. They encourage us to be patient, to accept the intangible as real, and to let our sixth sense guide us with its ancient wisdom.

7

Love Chemistry 101: Perfume and the Sixth Sense

Give me an ounce of civet, good apothecary, to sweeten my imagination.

—William Shakespeare, *King Lear*

In the 1997 movie *Batman and Robin,* actress Uma Thurman plays a deliciously evil character named Poison Ivy. The beautiful villainess has at her disposal an ingenious way to attract men: a glittery pink pheromone powder, which she calls her "love dust." In Ivy's deadly game of seduction, any man who sniffs this dust will fall into her irresistible trap.

In one scene, Poison Ivy enters a party displaying in her palm a cosmetic compact shaped like a poison ivy leaf. She opens the compact and blows the love dust into the crowd. The powder slowly winds its way in a tendril of pink smoke through the onlookers, drifting into their noses. In no time at all, the men in the crowd begin to swoon over Ivy; they have fallen in love with her without warning. But here's the catch: Any man who succumbs to Ivy's tempting "How about a little kiss?" will die the instant his lips touch hers.

That pheromones have been featured prominently in a Hollywood movie is not surprising, given that they are inherently mysterious and full of interesting potential, and this makes them promising candidates for starring roles. Does the existence of pheromones mean our best efforts to control ourselves, to keep our emotions in check, may be futile?

These naturally occurring molecules are intriguing because they invite us to think about ourselves in new and unexplored ways. They tempt us with tales of instant attraction, of chemical communication, of falling in love with someone based on chemistry instead of common sense. Tell us about our logical, thinking brains and we might react with some interest. But tantalize us with tales of love, lust, and sex fueled by chemical attraction and we sit up and take notice.

We have seen how the sixth sense affects the broad spectrum of our lives. Its voice is present in our decisions to plunge crazily into lust, to nurture our infants, to steer clear of a person or situation, to act out aggressively. The sixth sense and its anatomical link to the emotional core of the brain provide us with another sensory system that, in the end, makes us more feeling, more aware, more discerning, more *human*.

Assuming you have an intact vomeronasal organ, you live with your sixth sense every day. From the time you wake up to when you slip once again into sleep at the end of the day, your sixth sense is working busily, nonstop. Because it is programmed to affect you subconsciously, you may not even be aware of it until it issues forth a message or a warning that is impossible to ignore.

What if you could enhance your already industrious sixth sense by "feeding" it a substance that would make you feel better and more at ease with yourself and even more open to the intuitive capabilities of pheromone communication? Pheromones and the sixth sense have become the darlings of the perfume industry. Scent lovers can purchase perfumes and colognes that feature pheromones on the label and make claims ranging from the believable to the ridiculous.

Whether it comes in an expensive bottle from a boutique or finds form in a dab of simple rose water to the wrist, fragrance engages us in a subtle exchange between scent molecule and olfactory cell. If you're like most of us, you like to stimulate your olfactory system with fragrance, whether on your body or in your surroundings. And, you can probably remember odors from your childhood with amazing accuracy; as you do so, you open channels of memory long dormant. What could be sadder than a world without fragrance? Smells lend color to life. A bad mood can be lifted with scent, and a sullen living room or bedroom can be infused with energy. Advertisers tell us that a man can make himself feel more powerful by wearing a certain scent. A woman can do the same, or she can choose to be as soft and delicate as a kitten. Feeling playful and sexy? Young and innocent? Ready to take command of the corporate boardroom? All geared up for a sporting afternoon on the polo field? Are you a rogue today or a scholar? A lady or a tramp? Choose your fragrance and become who you want to be. All it takes, the marketing executives promise, is a splash of perfume or cologne.

Our love of scent has created an industry determined to fill, through "image advertising," every possible need, wish, and desire. Department store cosmetic counters display hundreds of fragrances, each clamoring to drape the wearer in a specific scent image. Indeed, we choose our fragrance so carefully because it makes seductive promises, and perfume ads tell us we can enhance our personalities simply by wearing the perfume or cologne in question. There are more than 1,200 fragrances on the market, each with a message and a mood, and the message appears to be working as hoped. In a survey conducted by the *New York Times* Marketing Research Department and the Olfactory Research Fund, six out of ten people said they believe the sense of smell can affect emotions. Sixty-three percent said scent could enhance the emotion of love, 61 percent said scent contributed to feelings of joy or happiness, and 57 percent associated certain odors with contentment.

These days, opening a fashion magazine usually becomes an unplanned journey into the world of scent: pockets of fragrance lie tucked within the pages, their molecules sneaking out to tickle your nostrils. If you pull the fragrance flap apart, the scent experience heightens. If you like the fragrance, that is even better because a successful perfume can earn many millions of dollars for the company that makes it, and perfumes are given hefty marketing budgets because the financial stakes are so high. One survey revealed that 55 percent of women use fragrances five to seven times a week. If a perfume manufacturer can convince a woman to be eternally faithful to his product, the future indeed looks bright for his bottom line.

And so it's no surprise that we want to smell good, to let the world know we care about our bodies and hygiene. We want our homes, offices, and social settings to be imbued with a fragrant reminder of ourselves. The smell of our own sweat lingering on our bodies is not an acceptable part of our society, as it is in some cultures, so we scrub ourselves clean and re-create ourselves with scent from a bottle. The advertisement for Hugo men's cologne makes a direct, and perhaps convincing, reference to our collective odor phobias: "The world is getting smaller. Smell better."

We are also led to believe that scent can cure a faltering love life. In her book *Love Magic,* modern-day apothecary Marina Medici lists nine scents that "will bring magical help to most of the issues which might arise in your love life." Write these down if you want to breathe new life into a romance: neroli, ylang-ylang, rose, jasmine, Melissa, Damiana, heather, a combination of apple and nutmeg, and ginseng. Coco Chanel suggested the following simple and elegant tactic: "A woman should wear perfume where she expects to be kissed." The French even have a saying, "A woman who wears no perfume has no future." On the other hand, if we were to follow the reasoning of Greek biographer Plutarch, we wouldn't need bottled perfumes. Many centuries ago he wrote, "A man in love is full of perfumes and sweet odors."

Fragrance: A Glimpse into the Past

As far back as 4500 B.C., the Egyptians used fragrance for a variety of purposes. Whether to enhance the sensory experience of their worship rituals, to cover up the stench of their embalming practices, or to erect perfume "shields" against evil spirits, the Egyptians were true connoisseurs of fragrance. Wealthy Egyptian women wore on their heads scented wax cones that would melt during the course of an evening's festivities and coat them in a sheen of exotic perfume. The most famous of the Egyptian fragrance hounds was Cleopatra, whose cedarwood boat floated on sails steeped in perfume. Cleopatra's body was a study in careful scent application: on her hands, *kyphi* (oil of crocus, rose, and violet), and on her feet, *aegyptium* (a lotion containing almond oil, henna, honey, cinnamon, and orange blossoms).

Perfume's origins have been traced to Mesopotamia, the ancient region of southwest Asia between the Tigris and Euphrates Rivers located in modern-day Iraq. Mesopotamia lacked sanitation and therefore had a horrible odor. One particularly foul-smelling ritual—the burning of sacrificial animals to appease the gods—produced an acrid smoke that hung stubbornly in the air. Only incense could cut through the curtain of odor. The Mesopotamians also burned incense after sexual intercourse, and they believed in the healing and evil-banning powers of scent.

Fragrance held a heavenly meaning for the ancient Greeks: Their mythology told them that gods and goddesses descending to earth always arrived in a cloud of scent. After the deities had reascended to their celestial homes, wildflowers would grow where their feet had walked. The Greeks perfumed their bodies according to a rigorous regimen of one scent per body part. The head would be anointed with rose or apple, while the arms would receive mint and the legs, wild ivy. Upper-class Greek women were scented in a most unusual way: Their slaves took the fragrance into their mouths and sprayed their mistresses' bodies with the perfume. Well-to-do Romans carpeted the floors of their homes in layers of rose petals, and the Roman emperor Heliogabalus bathed in water scented with an opulent rose wine.

That today's sophisticated perfumes trace their origin to incense is evident in the word's Latin roots: *per* (through) and *fumar* (to smoke). Many cultures still use incense as part of daily rituals, and lighting an incense stick is as natural as pouring a cup of morning coffee is to an American. In one well-known shop in the Nepalese city of Kathmandu, where tourists go to buy intricately painted wall hangings called *thankas*, the shop's proprietors are devotees of a variety of incense that stings the eyes and lays sandpaper across the throats of the uninitiated. Nevertheless, the goods are impressive enough to keep coughing and sneezing Westerners seated patiently as they view the paintings. In Kathmandu, where unsightly mounds of garbage and other detritus line the streets and creep through the alleys, incense greets the stink like a spirit of hope.

For hundreds of years, professional perfumers have been perfecting their craft, mixing scents and essences in complex combinations to achieve the desired result. The fragrances they work with include floral, musky, resinous, minty, and acrid. To properly smell a perfume, it must be applied to the skin or to a handkerchief. Sniffing it from the bottle overwhelms the nose, as the perfume's three notes (top, middle, and base) and the alcohol base can overload the olfactory cells.

The perfume industry's *modus operandi* is rooted in the fact that people remember and are affected deeply by smells. Scents have an unnerving capacity to stir recollections. It is as if the scent molecules themselves are embedded in our memories, so that when we think about the way a childhood backyard smelled—redolent of daffodils, raspberry bushes, lilac trees, and an old wooden rainwater barrel—we are in effect calling up each scent memory one by one as if reproducing it in a laboratory. Numerous studies have shown that people can recall smells with a high degree of accuracy, much higher, say, than they can recall visual images after time has passed.

We can send compelling messages through scent. Before director Paul Bern, the husband of actress Jean Harlow, killed himself, he doused his body in Harlow's favorite perfume, Mitsouko, a blend of oak moss and peach. When Bern was found dead, he was cloaked in a scent cloud of his wife.

Smells transcend our ability to express ourselves with words. Can you describe the way your lover smells? As you try to do so, you might assign words and phrases to the scent memory. But can you do verbal justice to the way he smells after lovemaking or the way her scent has become woven into the sheets and pillowcases? One woman told us her lover's body odor reminds her of her grandmother's curtains—comfortable, old, a bit dusty; however, this description doesn't truly describe the way this man smells. Words fail when we attempt to describe odors, just as they do when we try to put exact words to the chemical conversations instigated by pheromones.

Scenting Our Surroundings

What olfactory sensation registers in your nose when you walk into a bakery? Is it the timeless aroma of fresh bread? The cloying sweetness of glazed doughnuts? Bakeries seldom fail to elicit a powerful smell sensation because they are like olfactory theme parks, with a delightful smell in every corner.

Our noses have much to say about how we view and experience our surroundings. A house for sale that smells of fresh-baked bread or cinnamon sticks tends to sell more quickly, and smart Realtors prepare for a house showing by first giving it the pleasant smell of being lived in.

Smells enhance memories and boost our enjoyment of activities and events. The human nose can detect more than ten thousand odors and, if the olfactory cells have not been damaged, can identify each with remarkable precision. Florida's Walt Disney World knows that special times register more deeply in the brain's memory banks if they are accompanied by scents, and the park has added the pungent fragrance of burning wood to its Pirates of the Caribbean ride. At the PuroLand theme park near Tokyo, movie theaters are equipped with "odor chairs," each of which has a built-in system for delivering smells that correspond to what viewers are watching on the screen. And at the Mirage Hotel in Las Vegas, the atmosphere in "Polynesia" is enhanced with coconut room fragrance that streams from the building's ventilation system.

The Japanese have found that workers exposed to jasmine, lavender, and lemon fragrances seem happier and make fewer computer keyboard errors. Anxiety and other forms of stress can be treated with essential oils and other forms of aromatherapy. In the classroom, certain odors (eucalyptus and lemon oil, for example) can improve concentration and recall. Some patients undergoing magnetic resonance imaging (MRI) scans, in which the body is placed inside a metal imaging drum, become claustrophobic and emotionally stressed. Researchers at New York's Memorial Sloan-Kettering Cancer Center gave MRI patients a synthetic vanilla fragrance, a "comfort" aroma, to sniff during their scans; the "sniffers" came out of their exams feeling significantly less anxious. Perfumes with a prominent vanilla note are perennially top sellers, which is why Shalimar, created by Guerlain in 1925, continues to reap impressive profits. Industry pundits think vanilla is desirable because it makes us feel safe, at home, warm. Sniffing vanilla is like being in Mom's kitchen while she bakes vanilla-laced cookies or hands you a vanilla cream soda brimming with sweetness. In a world of uncertainty and unrest, vanilla assures us that everything is going to be all right.

The emerging practice of "environmental fragrancing" has led fragrance industry prognosticators to deem it the next great trend of the future. The home fragrance industry already enjoys some $950 million in annual sales, and the scenting of environments outside the home stands poised to make the big time. A nationwide sense of smell survey revealed that eight out of ten respondents use environmental fragrances—scented candles, room sprays, and potpourri—on a regular basis.

The powerful effects of scent have not been lost on retailers and businesspeople. Scent sampler Arcade, Inc., is pioneering ways to deliver specially designed fragrances to shoppers in bookstores, music stores, and restaurants. The goal is to boost the shopping and eating experiences of those who encounter the scent. Not too far into the future, scent will find its way into increasing numbers of factories and offices as employers look for ways to enhance productivity and employee morale.

Undoubtedly one of the most important skills of an airplane pilot is being able to maintain a high degree of concentration for extended periods of time. Mark Peltier, president of Minneapolis-based AromaSys, a leader in environmental fragrance technology, is working with the National Aviation and Transportation Center to pipe citrus and wintergreen essential oils into the cockpits of airplanes. Studies have shown that these particular scents increase and promote alertness. And imagine what fragrance could do for stressed-out air-traffic controllers: They could ostensibly inhale a scent to sharpen their alertness and calm their nerves.

Dr. Alan Hirsch, the neurological director for the Smell and Taste Treatment and Research Foundation in Chicago, found that shoppers are more likely to spend money when a fragrance is wafting through the store. Specifically, Hirsch and his colleagues looked at how scent affected people's attitudes toward Nike-brand athletic shoes. Potential Nike shoppers were divided into two rooms: one that contained only filtered air and one that was scented with a light floral fragrance. Those in the floral room expressed more interest in the shoes and were more eager to buy them than the shoppers in the unscented room. And the shoppers in the floral room were willing to pay more for the shoes in question—an average of $10.33 more per pair.

To test his findings in the real world while in Las Vegas for a conference, Hirsch decided to find out if fragrance could subliminally encourage people to put more money into slot machines. He conducted his study at the Las Vegas Hilton, pumping fragrance into one area of the casino. The odorized area of the casino saw a 45 percent increase in the amount of money fed into the slots.

Researchers in another study piped a fruity-floral scent and a spicy scent into a jewelry store and found that although patrons did not buy greater amounts of merchandise, they did linger longer with the goods. The fruity-floral scent made both men and women stay in the store longer, but the spicy scent affected only men. Odors have also been found to play a role in how long people study museum exhibits before moving on. Where scents are present, exhibits get more attention from museum-goers.

Perfume: From Civet to Science

Perfumers began infusing their concoctions with animal substances during the early years of perfumery because such ingredients were thought (like pheromones) to affect the sexual behavior of humans. The most commonly used were *civetone* (also called civet), a musky fluid secreted by the anal glands of the civet cat; *ambergris,* a waxy, grayish substance formed in the intestines of the sperm whale; *castoreum,* a territory-marking substance of the Russian and Canadian beaver; *muscone* (also called musk, from the Sanskrit word for testicle), a greasy red secretion produced in the glandular abdominal sac of the male musk deer; and pig pheromones.

In the sixteenth century, England saw the arrival of civet, ambergris, and musk from Italy. The characters in Shakespeare's plays rubbed their bodies with civet to perfume themselves. Henry VIII favored a scent made from ambergris, musk, and rose, and Queen Elizabeth I wore a heady perfume of rose and musk. The trendsetting Elizabeth is also credited with introducing the pomander to the higher echelons of society. Hers was a pungent clump of ambergris, musk, and civet held together with aromatic gums and shaped into a ball. And the hedonistic aristocrats of the Napoleonic empire spent a good deal of time (and money) slathering their fleshy bodies in musk. Were their sex lives amplified by this application of musk? Perhaps that's not the point; in any event, they must have enjoyed themselves.

Another popular fragrance, angel water, combined the scents of myrtle, rose, orange flower, ambergris, and musk, and was considered to be the "in" fragrance of the eighteenth century. Fashionable ladies of the day applied the perfume to their breasts, which were more accessible thanks to their low-cut dresses and push-up corsets.

In China, courtesans dined on food flavored heavily with musk. The idea was that when the women were squeezed and fondled during lovemaking, they would actually sweat the musk, thus revving up the intensity of the love act.

Even today, as one fragrance study revealed, the majority of perfumes and colognes in a selected sampling featured musk as a key ingredient. Of 400 perfumes and colognes for women, 340 contained musk, 156 contained civetone, and 26 contained castoreum. Of the 350 male fragrances examined, 328 contained musk, 21 contained civetone, and 47 contained castoreum. In both samplings, several fragrances also contained pig pheromones, which were used as a fixative to slow the rate of evaporation, thus making the scent last longer. Before science made it possible to produce synthetic equivalents of animal ingredients in laboratories, the methods used to obtain civet, castoreum, ambergris, and muscone harmed the animals and were certainly no picnic for the person doing the collecting.

Also, our olfactory cells are "programmed" to receive and process the scent of musk. A study conducted by International Flavors and Fragrances, Inc., in New York City found that musk can affect a woman's reproductive cycle. Women who sniffed musk perfumes and colognes ovulated more frequently and had shorter menstrual cycles. Nonetheless, pheromone researcher Clive Jennings-White warns: "While providing a scent may elicit a positive pleasant response, this should not be confused with a pheromone response."

The use of animal ingredients has survived the changing face of perfumery. But as we have already noted, pheromones are species-specific, which means an animal pheromone won't elicit a response in the human VNO. A more likely explanation for the lasting presence of animal substances is that they act as highly effective fixatives, or carriers, for a fragrance's main scent components. Civet, for example, contains a steroidal molecule that evaporates slowly, which means it can bind to the skin for extended periods of time and thus "hold" the scent of a perfume to the wearer.

If pheromones from a pig have no effect on the vomeronasal

organ of a Manhattan socialite or a housewife in Illinois, why do we continue to spend our money on perfumes and colognes that contain animal pheromones? "What do we do every day?" asks David Moran. "We wash off our pheromones. And what do we replace them with? Perfumes. What are those? Pheromones from other animals. Most perfumes are musk-based, and animal musk glands are full of pheromones. It's so ironic: We wash off our human pheromones and replace them with animal pheromones, which do not work on humans."

Sex in a Bottle

In the novel *Perfume,* by Patrick Süskind, a perfumer declares his desire to "create a scent that was not merely human, but superhuman, an angel's scent, so indescribably good and vital that whoever smelled it . . . would have to love him, the wearer of that scent." This is a near-Orwellian concept: What if we could control other people—their desires, their intentions, their moods—by perfuming our bodies with potions designed to affect the behavior of our fellow humans? That day has not yet arrived, but as you read on, you'll find that some products want you to feel as if you have put another person into a Vulcan mind-meld of exquisite passion without even so much as blinking an eye.

It should come as no surprise that the selling of perfume has traditionally revolved around the subject of sex. We love to see artistic magazine photographs of naked male midriffs and slinky women. Whether their message is blatantly sexual or subtly erotic, perfume makers know that sex—or the promise of it—sells.

But in a market that is crammed with fragrances, it takes something special to stand out from the crowd. To succeed, a perfume maker must stir the imagination of the consumer and offer something enticing and new. To that end, some perfume manufacturers have jumped onto the pheromone bandwagon, adding what they say are human pheromones to their perfumes and colognes and making bold claims about what their fragrances can do. Scents that purport to contain human pheromones are now widely available.

Because human pheromones can foster romantic feelings, they have won the attention of entrepreneurs attempting to bottle passion. In France, one perfume features "a molecular complex for maximal attraction" that supposedly can "hook" any unsuspecting person within a radius of thirty feet. Andron eau de cologne boldly says it can "create an intensive magnetic field between the sexes." Some fragrances (such as Realm, described below) are backed with millions of dollars of research and extensive science, while others seem to come out of left field with their robust exclamations about aphrodisiacal effects. Anyone contemplating buying a pheromone fragrance should investigate the company behind the product. Snake oil sells simply because people become enthralled by flashy, if inaccurate, statements.

Awakening the Sixth Sense

Can pheromones in perfumes activate the human vomeronasal organ and send a message to the hypothalamus, the seat of emotional expression and feeling? Is the sixth sense truly awakened when we wear a perfume or cologne containing human pheromones?

Human Pheromone Sciences, Inc. (formerly called the Erox Corporation), founded in 1989 by David Berliner and with headquarters in Fremont, California, says it has created the first perfumes and colognes to contain patented, synthesized human pheromones: Realm Women and Realm Men. One dose of either fragrance is said to contain the equivalent of one naked body's worth of real pheromones.

Human Pheromone Sciences was launched with this mission: To explore how human pheromones could be added to perfumes and cosmetics to kick the sixth sense into high gear. To that end, the company put nearly $2.3 million into the research and development of synthesized human pheromones. In 1991, their research scientists presented their findings at the International Symposium on Recent Advances in Mammalian Pheromone Research in Paris.

The Realm product line can be found in many major department stores and includes perfume, cologne, body lotion, body powder, bath products, and a roll-on fragrance called Realm Roulette. All are designed to elicit positive feelings in the wearer and improve his or her mood by instilling a sense of well-being. This is accomplished, the company says, by reaching beyond the sense of smell to the pheromone-receiving VNO and the sixth sense. Realm's trademarked slogan appropriately urges consumers to "Awaken Your Sixth Sense!"

Human Pheromone Sciences studies have shown that women and men react differently to certain pheromones, so Realm perfumes contain pheromones that affect the gender they are designed for. While both Realm versions contain a combination of female and male pheromones, the women's fragrance contains a greater percentage of synthesized male pheromone and the men's fragrance has a greater percentage of synthesized female pheromone.

Human Pheromone Sciences emphasizes that its products are not aphrodisiacs, and press material for Realm cautions: "Our fragrances may enhance sensuality but not sexuality." They do this by making the wearer feel good. A person who feels good about herself or himself is more likely to be in the mood for a little cuddle or a night on the town.

The marketing of Realm, says Human Pheromone Sciences CEO William Horgan, takes a decidedly different approach from traditional methods. While many perfume ads depict exciting or romantic "don't you want your life to be like this?" scenes, Realm's advertising focuses on the science behind the product and how it affects the human VNO and the corresponding sixth sense.

"We feel we are the fragrance that truly can deliver what the others can only promise," Horgan told us. "All the [other perfume] ads are illusion—ladies in floating gowns, ladies in swimming pools. It is all imagery and no science. We had to find a way to convey that this is science. That was the challenge. Realm is the first fragrance that awakens your sixth sense."

Says David Berliner, "We were born naked, and in the old days, when nobody wore too many clothes or used deodorants, chemical communication was quite open. Now we're covered completely. We're not communicating enough to like each other. Everybody wants a sex attractant. Realm is not a sexual attractant. It just makes you feel good."

Realm entered the marketplace in September 1993 under the guidance of Pierre de Champfleury, the former president of Yves Saint Laurent Parfums. In the summer of 1996, financial analyst Dean Witter placed Human Pheromone Sciences eighteenth on its list of the "top 200 stocks in order of attractiveness based on relative strength." Human Pheromone Sciences' 1997 year-end sales were expected to reach $20 million.

According to a Human Pheromone Sciences press release, women who have worn Realm describe feelings of ease and an openness to others and to their own sensuality. They also feel warmer, happier, and less negative. Men say they have more self-confidence while wearing the cologne, as well as an increased sense of comfort with themselves.

The company introduced another pheromone perfume for women in the spring of 1997. Called Inner Realm, this scent is lighter and fresher than the rich Oriental undertones of the original Realm. Inner Realm contains a patented human pheromone designed to be received by the female wearer's VNO. The fragrance's pheromone has been likened to a turbo engine that can rev up the wearer's personality by making her feel energetic and empowered. One Inner Realm wearer, a female neurosurgeon, started to display an uncharacteristically "bubbly" mood, her friend told us. "She tends to be kind of negative about things," he said. "But, with Inner Realm she became more positive, more funny, more upbeat."

Realm's synthesized pheromones were created in a laboratory at the University of Utah. There, research scientists extracted pheromone molecules from samples of human skin and created synthetic (and more potent) versions of the original molecules. Organic chemist Clive Jennings-White led the research, publishing his findings in a paper titled "Perfumery and the Sixth Sense." Jennings-White discovered that two substances extracted from human skin have significant effects on the VNO: estratetraenol and androstadienone. As would be expected, the male VNO is highly sensitive to estratetraenol (the female pheromone), while the female VNO shows marked activity in the presence of androstadienone (the male pheromone).

As part of his research, Jennings-White tested selected popular perfume ingredients to see if they would affect the human VNO. Using a miniprobe designed to reach the opening of the VNO, Jennings-White was able to determine which perfume additives stimulated it and which left it cold. The scientist puffed civetone and muscone into the VNOs of volunteers; the organ did not respond to either substance. However, when the human pheromones estratetraenol and androstadienone were delivered, the VNO reacted in a display of physiological activity. The results of Jennings-White's experiment do not necessarily mean the end of the line for civetone and muscone, as these substances do stimulate high levels of activity in human olfactory cells and are thus effective activators of the sense of smell. Jennings-White's tests also revealed that pig pheromones, which are popular perfume additives, "show no significant activity in the human VNO."

One of the challenges of creating a pheromone-based perfume is ensuring that its ingredients work in harmony, so that the olfactory components of the perfume do not overwhelm the pheromonal components and vice versa. To achieve this, the makers of Realm experimented with a variety of percentages and concentrations, eventually arriving at the formulas now on the market. (Human Pheromone Sciences would not divulge Realm's pheromone-to-fragrance ratios, citing trade-secret reasons.)

Jennings-White also analyzed the ingredients of a number of other popular perfumes and colognes now on the market. Lydia, a perfume for women, features a potent concentration of pig androstenol. Lydia's androstenol is present in concentrations five hundred times that of the androstadienone pheromone in Realm, leading Jennings-White to surmise that "it is not possible to compensate for a substance being inherently inactive in the human VNO by increasing the concentration." Jovan's Musk-2 also features pig androstenol as a key ingredient.

"Only products which contain human pheromones are active in the human VNO, and there is excellent correlation between the nature of the components and the activity of the finished products," Jennings-White writes. He concludes, "We are approaching a radical shift in the concept of perfumery. Henceforth, the design of a perfume should take into account stimulation of the long-neglected sixth sense, the vomeronasal system."

Snake Oil?

How many men could honestly deny being intrigued by this promise: "Makes women go from NO! NO! NO! to PLEASE! PLEASE! PLEASE!"

Such claims dominate an advertisement for Submit, a spray cologne for men by Image Tec of Lynbrook, New York. Submit says it will detonate "an explosion of sensual desire, like touching a match to gasoline! It produces an almost hypnotic effect—so intense, so compelling, it must be fulfilled . . . at once! When this super-stimulant spray was tested on humans at cocktail parties, singles bars, offices, and even by door-to-door salesmen—back came reports of a glorious paradise of non-stop pleasure."

Exhausted yet? Submit doesn't pull any punches in its ads. Its active ingredient is Pheromone Prime, described by its makers as "a natural scent extract that arouses the female libido . . . heightens her response to a fever pitch." Our best advice is: Buyer beware. However, Submit does give refunds if, after fourteen days, the cologne doesn't "turn her into your willing and eager slave for up to 8 straight hours at a time."

Another cologne for men that also claims to contain pheromones is The Scent, a product of Adrian & Co. of Orlando, Florida. The Scent aims to "attract women instantly . . . on a primordial level (animal subconscious)." Its manufacturers assure men that The Scent is "a perfectly legal sexual stimulant . . . cleverly masked in a men's cologne . . . that when unknowingly inhaled by any adult woman fires up the raw animal sex drive in every woman." Buyers of The Scent get a bonus copy of "How to Seduce Girls" with every purchase (although we wonder why this is necessary given that the cologne claims to do all the work). And if The Scent isn't enough to turn an ordinary man into an instant just-add-pheromones Casanova, Adrian & Co. also sells one-ounce bottles of Meltdown Massage Oil with Pheromones for $7.95.

Love's Bouquet, from Wonda Products of Brooklyn, New York, is billed as a "unisex" pheromone perfume—it is designed to be used by women and by men. In the product's ad, the selling of sex, or the promise of attraction, is once again evident: "We call it Love's Bouquet. You'll call it animal magnetism in a bottle."

Ecstasy claims to contain "scientifically formulated pheromones that act to heighten your sex appeal. How? Pheromones produce scent signals that create an intense magnetic reaction between the sexes. A feeling of attraction." And Beaux Gest for men and Bare Essence for women "utilize pheromonal technology . . . designed to enhance your own natural attraction." In a direct appeal to women, the company that markets the fragrance says, "Bare Essence for women will draw men toward you like bees to flowers . . . then take your pick!"

Other fragrances billed as containing human pheromones include Formula 88 from Australia; Don Juan for the "dominant male" ("Are you sexually invisible?" whispers the headline for the Don Juan Seduction Shop in Lincoln City, Oregon); and Falling in Love by the Philosophy cosmetics company of Tempe, Arizona. Curiously, Pheromone, a perfume created in 1980 by the Marilyn Miglin Institute of Chicago, makes no claims that it contains actual pheromones. Touted as "the world's most precious perfume," Pheromone carries a serious price tag: $500 an ounce.

One product that has attracted attention is Athena Pheromone 10:13, an odorless fragrance additive produced by the Athena Institute for Women's Wellness in Chester Springs, Pennsylvania. According to its creator, Winnifred Cutler, Ph.D., Athena Pheromone 10:13 is "a cosmetic fragrance additive to help a woman increase her own sexual attractiveness." The additive's key ingredients are synthesized human pheromones, Cutler claims. Purchasers of the purported pheromones are instructed to add them to their own favorite fragrance, mix well, and apply daily.

Commemorating Cutler's birthday (October 13), Athena Pheromone 10:13 has a male counterpart, Athena Pheromone 10X, which her company says contains "synthesized human male pheromones that are a chemical copy of the natural pheromones produced by the body of a sexually-active male in his mid-twenties."

This is only a selection of fragrances that supposedly contain pheromones. An Internet search using the keyword *pheromone* will lead you to even more. But be forewarned: Your eyes may not be ready for what you discover as you surf. Some Internet sites feature color photographs of topless and nearly undressed women and men in seductive postures—the postures, no doubt, that your mate will inexplicably assume when under the influence of the love potion in question. In many cases, the accompanying descriptions of the pheromone fragrances try to entice the reader into thinking he or she can lure the opposite sex into complete (but willing) sexual submission. There are endless promises of sexual prowess and power, and, of course, convenient ways to order the products on the Internet with a simple click of a button.

Pheromone perfumes present a new road into the vast world of scent. Whether they are carefully tested preparations that have combined the seemingly incongruous worlds of science and perfumery—or snake oils with false claims peppering their literature—there is no shortage of options for the consumer. As pheromones continue to cement their status in the everyday lexicon, we will see more and more products with the word *pheromone* on the label.

In the meantime, synthetic versions of human pheromones may soon blaze a new frontier in medicine. As we'll see in the next chapter, scientists continue to examine innovative ways of treating human ailments and disease, using the VNO as a conduit for pharmaceuticals.

8

The Future of Pheromones

Don't you see? She's clouded your mind! She's infected us with some kind of pheromone extract!

—Batman (George Clooney), speaking to Robin (Chris O'Donnell) in *Batman and Robin*

Despite our fascination with technology, finance, designer clothing, and expensive automobiles, we are still drawn toward topics that involve—what else?—ourselves. We especially love to talk about how we can enhance our love lives or why a certain love affair was passionate and almost too hot to handle and another short-lived and dull. We like to talk about our interactions with other people and how our own bodies and body chemistries might be influencing such interactions. We want to know how pheromones affect all areas of our lives—from the boardroom to the bedroom.

And what will the future bring? Will pheromones be forever relegated to discussions of why we fell head over heels in love with the football player in college but not the handsome engineering student who had a much more promising future, or will they someday have more practical applications? As research into this evolving field of science continues, the answers will undoubtedly surprise and fascinate us. Remarkable things are happening with pheromone research, and it's likely that these invisible, odorless molecules produced in the skin will become an important part of your life in new ways.

Might there someday exist a magical love potion capable of turning people into eager and willing love slaves? Possibly, but for now it's important to focus on what pheromones and their derivatives may be able to do if current research and clinical studies pan out as anticipated. How you incorporate pheromones into your life will depend on your individual circumstances and needs. We do know this: Pheromone technology is poised to enter your local pharmacy and, potentially, your own medicine cabinet. And that, as you are about to find out, is very good news.

Paying Through the Nose

A mail-order cosmetics company in Australia purports it may have found the solution to dealing with people who don't pay their bills on time. The answer? Overdue notices sprayed with what the company claims are pheromones.

"Payment-inducing pheromones" are marketed by David Craddock of London-based Bodywise Ltd. The company has created a substance called Aeolus 7+, a blend of androstenol and androsterone that is said to put debtors in friendly moods, the hoped-for result of which is payment of the late bill. The pheromones in question are supposed to subconsciously convince the delinquent customer to pay the bill by making him feel more at ease and also more likely to get out the checkbook.

To find out if Aeolus 7+ could deliver on its promise, the company conducted an experiment in which it mailed one thousand overdue bills, half of which had been treated with putative pheromones. Though not spectacular, the results were interesting: Customers who received the treated bills remitted their payments 17 percent more frequently than did the customers who received the untreated bills.

Bodywise claims that the pheromones in Aeolus 7+ work on people's moods, but some scientists have said that it contains pig pheromones, which will not affect the human VNO. Nonetheless, Bodywise is dipping its toes into what could become a practice among retailers. Some human pheromones can help put people in relaxed, content moods. Perhaps there is a correlation to feeling at ease and spending money, whether on overdue bills or new merchandise.

What would happen if you could send authentic human pheromonal messages through the mail—pheromones that activate the VNO of the recipient and cause his or her hypothalamus, the seat of emotional expression and mood control, to respond accordingly? What if retailers could purchase a specially synthesized pheromone designed to make a normally conservative, unimpulsive person an overnight catalog shopaholic? Think about the possibility of opening a fashion magazine and being overwhelmed by the urge to buy the items advertised within.

What if you could be convinced at a subconscious level to pay your overdue bills, shop till you drop, or linger far too long in a department store due to the effect of a pheromone being piped into your environment? Could there come a day when purchase displays are equipped with hidden pheromone mists formulated to attract passersby and propel them to buy things they might otherwise ignore? And what if organizations soliciting money through direct-mail campaigns could imbue their literature with pheromones that would put you in a warm and fuzzy mood, making you more likely to contribute to the cause?

No, you think, *I could resist such appeals to my emotions.* Maybe. But remember that pheromones don't appeal to your logical side. Pheromones zip straight to the core of your reptilian brain, skirting your developed "thinking" cortex and moving to the region that controls your most basic inclinations and physiological workings. So, it's not entirely absurd to posit that pheromones could someday be used as sales tools in the never-ending competition to attract customers.

A Spoonful of . . . Pheromones?

Pharmaceuticals are big business, and the most successful companies earn billions of dollars every year. Success relies on the creation and marketing of drugs that capture broad-based consumer and physician attention—and loyalty. "Wonder drugs" are not easy to come by; only about one in thirty-five thousand formulas enjoys success in the marketplace.

The bane of pharmaceutical companies is the expired drug patent. An expired patent on a drug that was originally the property of one company means that any company can now make a generic version of it. Knowing that huge profits can be realized with the introduction of a fantastic new medicine and knowing that patents have time limits, pharmaceutical companies are always looking for new and innovative ways to treat the many ailments that can afflict the human body. The winners in the race to bring new drugs to people are those companies that go beyond the conventional and venture into unexplored territory.

Medical technology employs a variety of drug delivery methods, with pills, suppositories, and injections being the most common. Now nasal sprays and skin patches are also being used to deliver medicines. But what if a company were to discover a new method of drug delivery? What if a company could harness the powers of pheromones in drug technology? That company would be destined for success if it could show that the drugs worked with more efficacy than anything else available. Although the concept sounds futuristic, it is being explored now at Pherin Pharmaceuticals, a privately held biotechnology company based in Menlo Park, California.

Pherin Pharmaceuticals, Inc., was founded by pheromone researcher David Berliner in 1991. The company's main thrust is the research and development of pheromone-based pharmaceuticals, and it is the first to study how the relationship between the VNO and the hypothalamus could be used to treat illnesses. To that end, Pherin has spent millions of dollars investigating the very complex subjects of chemosensory neurobiology, neuroanatomy, and pheromone technology. Pending approvals from the U.S. Food and Drug Administration, Pherin hopes to introduce in the next few years a unique class of drugs that are administered through the nose—and, more specifically, through the vomeronasal organ.

This development could have additional benefits. When you take a pill, your body has to work to process the pill and separate its active ingredients from its inactive ones. Your internal organs become involved and the drug passes into your bloodstream, which means side effects can occur. Taking medicine can be an unpleasant experience. One goal of pharmaceutical companies is to create drugs that not only work well but also produce the fewest side effects. Drug companies know physicians will prescribe medicines that are easy on their patients.

Enter Pherin and its team of scientists and physicians who are creating *vomeropherins*—potent synthesized modifications of naturally occurring human pheromones—in a laboratory at the University of Utah. Vomeropherins are substances that have a physiological or pharmacological effect on the human VNO. These patented compounds elicit changes in the hypothalamus and may be able to affect a number of bodily functions that are controlled by this region of the brain. Basing its experiments on how the human VNO-pheromone system works, Pherin is developing pharmaceuticals that can be sniffed and then are transmitted through the nervous system to the hypothalamus. The company has already produced and has patents for more than 1,000 vomeropherins.

Each vomeropherin binds to specific chemosensory receptors located on the surface of the VNO. Like putting blocks through holes, if a square finds a square, a match is made; a square block can't go into a round hole. Once at the receptors, the vomeropherins create electrical impulses that travel along the neural pathways leading to the hypothalamus. The hypothalamus reacts and responds to messages to create changes in behavior or endocrine activity, depending on the molecular structure of the vomeropherin being introduced: improve mood, alter hormone levels in the blood, stop eating, calm down, increase sexual excitement. These findings are significant because they suggest a direct connection between the human VNO and the brain. Says pheromone researcher Louis Monti-Bloch, "We are meeting the requirements that have been used in VNO studies in other mammals: behavioral and endocrine changes."

Because Pherin's vomeropherins are not ingested, they do not have any systemic involvement. In other words, organs such as the stomach, intestines, and liver do not have to process these compounds before they are delivered to the brain to do their work. This will be the true virtue of vomeropherins: They only have to be sniffed in extremely small doses to exert their effects on the hypothalamus. Traditional therapies rely on circulating the drug through the bloodstream in order to reach the drug's target zone in the body. Vomeropherin therapy would directly affect the brain by stimulating key afferent nerves. While most medicines in pill form are prescribed in milligram quantities, a vomeropherin works with a fraction of that amount—a picogram, which is a millionth of a millionth of a gram. Explains David Berliner, "If you compare a picogram to a gram, it's like the length of a pencil eraser compared to the length from here to the moon." Also significant is the speed at which vomeropherins travel to the brain: One ten-thousandth of a second is all it takes for the molecule to activate the hypothalamus.

To test their hypothesis that vomeropherins affect brain activity, Pherin scientists conducted a series of double-blind, placebo-controlled experiments involving a number of their synthesized substances, some more potent than others and each with its own unique molecular structure. The results, published in the June 1996 issue of the *Journal of Steroid Biochemistry and Molecular Biology*, reveal that vomeropherins indeed cause hypothalamic activity, and certain vomeropherins were found to alter hormone levels in the blood. Hormonal changes occurred even when only a few molecules of vomeropherin were puffed into the nasal cavity. The vomeropherins, Pherin is pleased to report, "deliver instant neural messages to the brain."

Pherin is working with the FDA to develop a series of new "VNO-hypothalamus" drugs. The company thinks it may be able to treat a wide range of health concerns that have a hormonal component. Currently under development are vomeropherins for prostate cancer, breast cancer, panic disorders, and premenstrual syndrome.

The hypothalamus is also the command center of the emotions. It controls mood, anxiety levels, states of fear or calm; it also tells us if we're hungry or full, drowsy or alert, sexually aroused or not. Pherin hopes to someday introduce vomeropherins to treat acute panic attacks, anxiety, and phobias, as well as compounds that help people lose weight. By using vomeropherins to communicate with the hypothalamus, Pherin could ostensibly go even further—speed up or slow down a person's metabolism, ignite a flagging sex drive, improve sleep patterns, cure insomnia. The possibilities are mind-boggling.

Vomeropherins and Hormones: Exciting New Inroads

Pheromone research has led to a number of remarkable and significant scientific findings. For example, scientists have discovered that vomeropherins not only affect a person's mood but also drive changes in the endocrine system, which is governed by the hypothalamus. Pherin researchers have also found that vomeropherins can cause an increase in alpha waves in the cortical region of the brain. Alpha waves are those related to feelings of relaxation. Other vomeropherins do just the opposite: They increase the incidence of beta waves, which reflect a state of mental alertness. Yet other vomeropherins can decrease the rate of respiration, which also leads to a state of relaxation. Research has also shown that muscle activity is affected by certain vomeropherins: Some cause muscles to tense while others relax muscle fibers.

"As a biochemist working on control delivery systems for drugs, I think it's extraordinary," says Dr. Sergio Nacht, a biochemist and physiologist at Advanced Polymer Systems, a San Francisco–based biotechnology company currently working on new drug delivery systems. "I foresee dozens, if not hundreds, of applications for this technology. It could be used to control different aspects of human behavior, the physiology of the human body, and mood and attitude. What is most amazing of all is the notion of controlling all that just by inserting a few molecules into the nasal cavity. I would say it is a revolutionary discovery."

When asked how pheromone technology and research might progress, Louis Monti-Bloch is enthusiastic. For example, he thinks vomeropherins could one day be used to treat male sex offenders, who have been said to possess abnormal levels of testosterone in their blood. A vomeropherin designed to alter testosterone levels would direct a message to the hypothalamus to alter the production of gonadal hormones. This in turn would control production of testosterone in the testes and could be used as an effective treatment in the future.

Hormonally Yours

In mammals, some pheromones have been shown to alter hormone levels and affect the animal's fertility or sexual behavior. Because the human VNO is thought to be linked by neural connections to the hypothalamus, vomeropherins might be administered to create hormonal changes in humans as well, as we have just described.

The hypothalamus, working in conjunction with the pituitary gland, secretes "releasing factors" that control the release of hormones in the body, including luteinizing hormone-releasing hormone (LHRH), a releasing factor secreted by the hypothalamus, and follicular-stimulating hormone (FSH), a hormone secreted by the pituitary gland, both of which are necessary components of the human reproductive system. LHRH is significant because its target is the pituitary gland, which is linked to the hypothalamus by a special blood circuit in the brain. When the pituitary is stimulated by LHRH, it releases luteinizing hormone into the bloodstream. Luteinizing hormone travels to the testes in men and to the ovaries in women, which when stimulated by the hormone produce sex hormones (testosterone in men and estrogen in women).

Pherin scientists have found that certain vomeropherins can alter blood levels of LH and FSH. Of particular interest to Pherin, however, is that a certain vomeropherin can decrease testosterone levels in the bloodstream of males. The ability to reduce or boost levels of circulating testosterone and other hormones just by puffing a vomeropherin into the nose is a major development in the world of medicine. Clinically, this discovery could mean dramatic advances in the treatment of cancers, some of which, including cancer of the prostate gland, feed on testosterone and need the hormone in order to grow. In women, estrogen has been implicated in breast cancer. Estrogen is also controlled by the hypothalamus-pituitary-LH link.

In addition, vomeropherins could open doors in the arena of hormone replacement therapy for men and women. Rather than administer replacement hormones through pills, skin patches, or injections, Pherin's pheromone technology could normalize hormone levels with vomeropherins, thereby minimizing the side effects associated with conventional drug delivery systems.

In some cases, testosterone levels in men could be raised to treat hypogonadism or testosterone deficiency, the latter of which can be caused by old age, alcoholism, testicular disease, or Klinefelter's syndrome. Men who have low testosterone levels sometimes experience fatigue, weakness of the muscles and bones, depression, diminished libido, and impotence. According to the Testosterone Source, an Internet site sponsored by SmithKline Beecham Pharmaceuticals and which provides information on how testosterone levels can affect men in middle age, testosterone deficiency can also result in osteoporosis and muscle wasting. One of the Web site's consultants, Dr. Adrian Dobs of Johns Hopkins University Medical School, estimates that five million American men do not produce enough testosterone and only about 5 percent of them seek testosterone replacement therapy.

Birth Control

Women are fortunate to have access to a variety of birth control methods, including condoms, cervical caps, IUDs, diaphragms, spermicides, and drugs that control conception by altering hormone levels (the Pill, Depo-Provera, Norplant). Oral contraceptives are extremely popular. Some ten million American women use them—that translates into about 60 percent of women between the ages of fifteen and forty-five. In the United States alone, annual sales of birth control pills and other hormone-based contraceptives reach into the billions of dollars.

Contraceptives containing synthetic hormones work by preventing ovulation. These hormones communicate with the hypothalamus and the pituitary, telling them to halt the production of hormones that stimulate ovulation. Because fertility is dependent on correct hormonal balances and production levels in the body, and because those physiological events are controlled by the hypothalamus, vomeropherins could one day be used as contraceptives instead of pills. Pherin is exploring this possibility, as the worldwide market for birth control is expected to grow rapidly. A vomeropherin designed to prevent pregnancy would indeed have global appeal. The drug would be easy to take, have few or no side effects, be effective in tiny doses, not leak into the bloodstream, and work quickly. And, as David Berliner points out, vomeropherins could also act as "morning after" contraceptives.

Prostate Cancer

Prostate cancer is the second leading cause of death among American men. In 1996, 317,100 cases were diagnosed and 41,400 men died from the disease. Those at risk of prostate cancer are men over the age of fifty and, possibly, those who eat a high-fat diet. The first signs of prostate cancer are a frequent need to urinate, painful urination, pain in the back or pelvis, and weak urine flow.

The current recommendation is that all men over the age of

fifty have a digital rectal exam every year. Abnormalities in the prostate can be detected in the early stages of disease with the use of a prostate-specific antigen (PSA) blood test. Often, the prostate gland is removed (prostatectomy) upon discovery of abnormalities or cancerous cells. Other treatments include radiotherapy and the prescription of drugs that prevent cancerous cells from feeding on hormones.

Prostate cancer can be treated with a number of drugs in pill form that reduce testosterone in the body. They include Lupron Depot, a product of TAP Pharmaceuticals, and Zoladex, a product of Zeneca Pharmaceuticals. Combined sales of these drugs, the dominant therapies for testosterone reduction, reach into the hundreds of millions of dollars each year.

Vomeropherins may someday provide an alternative treatment for prostate cancer. Pherin scientists are investigating a vomeropherin that can decrease a man's testosterone levels and, by design, slow the growth of a tumor in the prostate. And a vomeropherin would produce fewer side effects than do the standard therapies.

Feeling Anxious? Sniff This . . .

You are about to board an airplane. The skies in the distance look tumultuous and the air is crackling with lightning. *This,* you think to yourself, *is not going to be a very pleasant flight.* You feel a sudden wave of panic. Your palms begin to sweat. Your heart drums inside your chest. Your mouth is dry. As the plane taxis and prepares for takeoff, you reach into your carry-on luggage for a small container that resembles a bottle of eyedrops. You uncap the bottle and inhale the substance into each nostril. Instantly, you feel relaxed. The effect lasts for a few hours and then gradually wears off. You risk no addiction to this substance and will experience no side effects. A miracle drug? Perhaps. The substance described in this hypothetical scenario is a vomeropherin designed to put the lid on acute (short-term) anxiety.

Anxiety is generally marked by tension, feelings of apprehension and danger, and an overall sense of unease. According to the American Academy of Family Physicians (AAFP), one category of anxiety "can be a normal 'alarm system' alerting you to danger." But, the AAFP adds, "sometimes anxiety may go out of control, giving you an overwhelming sense of dread and fear for no apparent reason. This kind of anxiety can disrupt your life."

Panic disorder is a form of anxiety in which people experience repeated periods of extreme panic. The person suffering a panic attack will feel tightness in the chest, dizziness, and faintness, and might even begin to choke. He or she will also experience sweating, a feeling of being smothered, nausea, trembling, and perhaps hot flashes or chills. The heart will start to beat faster. The attack can last from five minutes to half an hour, depending on its severity. According to the National Institute of Mental Health, three million Americans will have a panic attack during their lives.

Anxiety can be treated with a class of anxiolytic drugs called benzodiazepines, which can sedate the taker as well as provide hypnotic, anticonvulsant, and muscle-relaxing actions. (Valium is one example.) Before benzodiazepines were introduced, the standard treatment was to prescribe barbiturates, which are serious sedatives with a cadre of side effects. Benzodiazepines, which work quickly, give the person taking them a feeling of euphoria; however, they also produce side effects, among them memory problems and impaired balance. Benzodiazepines can also be addictive, especially when they are taken with alcohol.

Another class of anxiolytics is azaspirodecanediones, one of which is the drug buspirone, a product of Bristol-Myers Squibb. Buspirone is less sedating than benzodiazepines and is less addictive when alcohol use is involved. The main difference is that buspirone does not create a sense of euphoria. Thus, doctors can find it difficult to wean their patients off the euphoria-producing benzodiazepines and switch them to buspirone. The best antianxiety drug, understandably, would have a quick onset and produce minimal side effects.

Anxiolytics generate about $2.5 billion in annual worldwide sales, with U.S. sales accounting for 56 percent of that total. Other major markets for antianxiety drugs are Italy, France, Germany, and Japan. According to Pherin, "The anxiolytic that will capture a major market share in the future is the drug that acts immediately, does not cause dependence, does not cause sedation or drowsiness, limits the possibility of overdose, does not interact with alcohol and is, most importantly, effective in avoiding the withdrawal symptoms of the benzodiazepines."

If pheromone pharmaceutical research goes as planned, people might someday be able to sniff vomeropherins and eradicate any episode of acute anxiety, whether while aboard an airplane or at an important business meeting. When sniffed into the nose, the vomeropherins would travel to the VNO and, once there, send a signal to the brain to replace anxiety with a feeling of calm. Such a treatment is not yet available at your local pharmacy, but Pherin scientists are now engaged in clinical trials of these substances in the hope that they can soon bring them to the marketplace and, in the process, "truly revolutionize anxiolytic therapy."

The Future of Weight Loss

Pound by pound, Americans weigh more with each passing year. In fact, statistics indicate that nearly one-third of Americans are obese. Ironically, even as we put on weight we are always on the lookout for ways to take it off. That we are a nation obsessed with thinness is overshadowed by our actual measurements.

Proper diet and exercise, the best ways to slim down, are not very appealing options to those who would rather turn to drugs to rein in their appetites. That so many people are fat is good news for companies peddling diet drugs. Americans' efforts to lose weight are staggering. Diet doctor Stephen Gullo tells us in *Thin Tastes Better* that Americans spend $33 billion each year in the quest to be skinnier.

Pharmaceutical companies are aware of the widespread desire to lose weight and some have responded with drugs specially formulated to help people become thinner. Recent attention has been focused on drugs that control appetite by monitoring neurochemicals, specifically serotonin, in the brain. While it's the stomach that signals the physical sensation of fullness or emptiness, it's the hypothalamus that talks to the stomach, sending messages that compel us to eat when hungry or stop eating when full. In fact, studies have found the hypothalamus possesses a sophisticated method of indicating hunger or fullness. An interesting theory about the hypothalamus is that it is programmed to maintain an individual's body weight at a specific level; such maintenance may be related to the quantity of fat cells present in a person's body at birth.

The latest generation of weight-loss drugs is significantly different from the amphetamines that entered the weight-loss market in the 1960s and which have since lost favor because they are so highly addictive. In April 1996, the antiobesity drug Redux (dexfenfluramine) obtained FDA approval and entered the market. The drug raises the brain's serotonin levels, thus controlling the urge to nibble. In short, Redux refocuses the taker's thoughts away from food, putting a stop to out-of-control eating and bingeing. Redux quickly became the darling of the diet drug world, and from June 1996 to April 1997, 140,000 doctors wrote 3.3 million prescriptions for it. Despite its commercial success, Redux began to receive criticism because of its side effects, which include headaches, dry mouth, diarrhea, and, in some cases, primary pulmonary hypertension.

Another popular diet-drug regimen, a combination of fenfluramine and phentermine called fen-phen, also enjoyed financial success. Fenfluramine raises the brain's serotonin levels but can cause grogginess, which is why it is prescribed with phentermine, a stimulant. Like Redux, fenfluramine has been implicated in some cases of primary pulmonary hypertension, and people taking fen-phen have also reported side effects that include insomnia, rapid heartbeat, and short-term memory loss. In September 1997, both Redux and fenfluramine were recalled from the market after the FDA found that use of the drugs may have been connected to serious heart problems in some people. At this writing, however, phentermine was still available for prescription.

The pharmaceutical market would understandably welcome an effective diet drug that has no serious side effects. To that end, Pherin is investigating how vomeropherins could control a person's appetite by telling the hypothalamus to dull the urge to eat. Because a vomeropherin appetite-control drug would not involve the body systemically, it would not produce any measurable side effects.

Controlling Pests with Pheromones

Another branch of research investigates how pheromones can be used to control pests. There are a variety of methods employed in this evolving science, but in general a pheromone can be used to either attract the pest to a bait (such as sticky paper) that will kill it or disrupt the pest's mating patterns and thus limit its ability to reproduce.

Judging from the continuing debate over the safety of pesticides, pheromone-based pest-control techniques will continue to be studied and developed. The following examples describe how pheromones are being used to control two ubiquitous pests: the cockroach and the stem-borer caterpillar.

Seduction of a Cockroach

Cockroaches can be controlled with synthesized versions of their own pheromones. Cornell University chemist Dr. Jerrold Meinwald and his team of researchers duplicated the molecular structure of the female cockroach's mating pheromone, and by doing so created an elixir that's irresistible to the male. When the male's pheromone receptors lock onto the presence of what appears to be the real thing, he follows the magnetic pull of his sensors toward the pheromone, which is detectable in quantities of just one-billionth of a gram. The love-struck male literally walks into a trap in his search for the female. It is then possible to pinpoint the placement of insecticides rather than spreading them over a broad area. Some scientists envision the cockroach traps containing "cockroach pathogens" that the lured males would unknowingly transport back to their colonies. As if that weren't enough, the males' mating behaviors are also likely to be severely scrambled; therefore, the numbers of their offspring would be diminished.

Controlling cockroaches with pheromones appears to be not only effective but also safe. This is good news for hospitals and schools, where the use of large quantities of pesticides poses obvious dangers.

Sensory Overdrive

Scientists working at England's Natural Resources Institute have isolated and synthesized the mating pheromone of the female stem-borer, a caterpillar that has decimated thousands of acres of valuable rice crops around the world. While applications of standard pesticides have done little to prevent the caterpillar from eating food intended for human consumption, the scientists discovered that the pheromone released by the female when it is ready to mate at its adult moth stage of life can be used to draw in the male of the species and thus disrupt the moth's mating patterns.

When a male moth senses the real or synthesized pheromone of a female, he tries to find it. But when large quantities of synthesized pheromones are released throughout a broad area, the male is unable to determine the signal's point of origin. He becomes confused: Where is this "female" who so recently announced her intent to mate? As the male's sensory system shifts into overdrive, he will begin to circle and flap himself into a heightened state of agitation. Some males will even flap themselves to death. In any event, because the males run themselves ragged or die before they reach the females, there are subsequently fewer eggs laid onto the stems of the rice, which means there are fewer moth offspring to contend with.

Stay Tuned

Pheromone research continues to heat up. Stanford University in California has delved into this science by showing that functional magnetic resonance imaging (MRI) scans of the brain may be able to show the activation relationship between vomeropherins and the hypothalamus, in what researchers refer to as "localization and identification of the neural substrates underlying pheromone processing in the human brain." The Stanford scientists discussed their work at the 1997 International Symposium on Olfaction and Taste held in San Diego, California.

Also of note is the fact that pheromone research continues to garner attention and interest at scientific conferences on human physiology. Pheromones were featured prominently at the Thirteenth International Symposium of the *Journal of Steroid Biochemistry and Molecular Biology* held in Monaco in May 1997.

As this book was in production, Pherin Pharmaceuticals signed a multimillion-dollar contract with the pharmaceutical division of the Dutch chemical company Akzo Nobel NV. NV Organon, Akzo's pharmaceutical arm, will work with Pherin to develop a series of vomeropherin drugs. Exactly which drugs are in line for development has not yet been disclosed. Although the terms of the agreement are being kept under wraps, it is known that Organon's main interests lie in birth control pills and other drugs that affect the human reproductive system, as well as psychiatric medications.

That Pherin has secured a substantial financial commitment from one of the world's leading pharmaceutical companies (the estimated cost to bring a vomeropherin compound through the FDA and into the consumer market is more than $200 million) carries enormous implications for the future of medicine. One Pherin investor has called vomeropherin technology a "quantum leap" in pharmaceutical product development.

When vomeropherin pharmaceuticals hit the market, be prepared for a storm of activity. Because vomeropherins are unique, and so different from anything now available, they will be the focus of intense publicity and perhaps even controversy. Maybe one day you'll use vomeropherins to treat a case of jangly nerves or a serious medical condition.

Epilogue

Rely on your sixth sense, that little voice within.

—Zimbabwean novelist J. Nozipo Maraire

As this book was nearing completion, one of us was asked a provocative question. The inquiry, simple and pointed, had a profound effect: "So, why should anyone care about pheromones?"

This is a good question. Aside from their potential pharmaceutical developments, why should we care about invisible, odorless chemicals produced in our skin and floating through the air? Does it really matter that we take the time to learn about what is happening in our lives at a subconscious level? Besides, if something is subconscious, we can't control it, so why worry, why bother?

The answer to these questions lies in the age-old human quest to know as much about ourselves as we can. We gobble up books that tells us about our lives, relationships, looks, careers, friends, homes, vacations, cars. Pheromones, then, give us another tool of understanding, another lantern to guide us through the vast land of our humanity. As complex creatures, we are eager to learn more about our complexity.

Our hope is that this book has given you a clearer picture of what pheromones can and cannot do and what they mean to you. Someday, you might even fill a prescription for a vomeropherin designed to treat your health concerns. In any event, now that you know about pheromones, we believe you'll never again look at yourself or the people in your life in quite the same way. You may know someone who made you feel uneasy at first meeting. Before learning about pheromones, you would have described the encounter as uncomfortable because you didn't know what to say to this person, or felt that he or she just didn't like you. Now, it may slowly dawn on you: Did the encounter go sour because of "bad" pheromones? Perhaps you'll meet someone and feel an instant bond. You'll think of pheromones and how they act on your subconscious mind to tell you crucial information about other people.

Pheromones help you round out the picture of your life. They fill in the blanks and do their part to help you in many situations. Enjoy your future chemical communications and, by all means, pay attention to them!

NOTES

We have avoided using numbered footnotes in the text because we intend this book for a popular audience and have thus kept away from too much of an academic or scientific slant. However, we feel it is important that you know our sources and, should you wish to read more about human pheromones and their related topics, have access to the articles and textbooks we used in our research.

Chapter 1. Chemical Conversations

- Diane Ackerman's book *A Natural History of the Senses* (Random House, 1990) explores the senses of smell, touch, taste, hearing, and vision.
- Dr. William C. Agosta's book *Chemical Communication: The Language of Pheromones* (Scientific American Library, 1992) focuses on pheromones in the insect and animal worlds but includes a chapter on human pheromones. The story about the dancing peasant and his fragrant handkerchief is from the chapter "Human Attractions." Dr. Agosta

and his colleagues isolated and determined the structure of the attractant and aphrodisiac pheromones of the Syrian golden hamster.

- The passage that describes the unusual odor preferences of France's Louis XIV are detailed in the article "Get a Whiff of This," by Richard Klein, *The New Republic,* February 6, 1995.
- The sentence that begins "Journalist Bill Moyers once asked mythology scholar Joseph Campbell the question . . ." refers to the book *The Power of Myth* (Doubleday, 1988), by Joseph Campbell with Bill Moyers that describes in detail the evolution of romantic love and the myths surrounding love and romance.
- *The Alchemy of Love and Lust: Discovering Our Sex Hormones and How They Determine Who We Love, When We Love, and How Often We Love* is by Theresa L. Crenshaw, M.D. (G. P. Putnam's Sons, 1996). Crenshaw, a physician and sex therapist, has conducted numerous studies that delve into the intricate world of the chemicals that instigate and dictate sexual interest, aggression, and passion. Additional information about the hormone DHEA can be found in numerous magazines and newspapers. For example, the June 1996 issue of *Vanity Fair* contains an article by Gail Sheehy titled "Endless Youth," which discusses the use of DHEA to "reverse" the aging process. *Time* (November 25, 1996) features an article by Jeffrey Kluger titled "Can We Stay Young?"; it investigates various anti-aging methods, including the use of supplemental DHEA.
- The information on the discovery of the human VNO was gleaned from interviews with Dr. David T. Moran and from the chapter "The Structure and Function of the Human Vomeronasal Organ," by D. T. Moran, L. J. Stensaas, L. Monti-Bloch, and D. L. Berliner, in *Handbook of Olfaction and Gustation,* ed. R. L. Doty (Marcel Dekker, Inc., 1995, pp. 793–820).
- A discussion of the anatomy of the vomeronasal organ by D. T. Moran, B. W. Jafek, & J. C. Rowley is presented in the research paper "The Vomeronasal (Jacobson's) Organ in Man: Ultrastructure and Frequency of Occurrence" in *Journal of Steroid Biochemistry and Molecular Biology* 39(4B):545–52 (Pergamon Press, 1991). This edition of the journal compiles information from the proceedings of the International Symposium on Recent Advances in Mammalian Pheromone Research, which took place in Paris, France, on October 6–9, 1991.
- The paragraph beginning "The hypothalamus, often called the 'brain's brain' . . ." provides a brief description of the function of the hypothalamus. For a more detailed discussion of what scientists have for years called the primitive brain, see *Origins of Neuroscience: A History of Explorations into Brain Function,* by Stanley Finger (Oxford University Press, Inc., 1994, p. 284).

Chapter 2. Pig Breath and Other Animal Wonders

- "No wonder the poor, confused female truffle pig . . ." refers to descriptions of truffles and why pigs find them so irresistible as presented in *Shadows of Forgotten Ancestors,* by Carl Sagan and Ann Druyan (Ballantine Books, 1992), and *Scent: The Mysterious and Essential Powers of Smell,* by Annick Le Guérer (Turtle Bay Books, 1992).
- Our discussion of the mammalian vomeronasal organ was augmented with information provided by pheromone researcher Dr. David Berliner and other pheromone scientists.
- For general information on the role of pheromones in the plant and animal worlds, see "Chemical Ecology," by P. H. Abelson, *Science* Vol. 264, No. 5158, p. 487, April 22, 1994 and "Olfactory Recognition: A Simple Memory System," by P. Brennan, H. Kaba, and E. B. Keverne, *Science* Vol. 250, No. 4985, p. 1223, November 30, 1990.
- The section titled "The Johns Effect" refers to the work of Dr. Margaret A. Johns, who published the following findings: "Urine-induced Reflex Ovulation in Anovulatory Rats May Be a Vomeronasal Effect," *Nature* Vol. 272, No. 5652, pp. 446–48, March 30, 1978 (paper coauthored with H. H. Feder, B. R. Komisaruk, and A. D. Mayer); "The Role of the Vomeronasal System in Mammalian Reproductive Physiology," in *Chemical Signals,* ed. Dietland Muller-Schwarze and Robert M. Silverstein (Plenum Publishing, 1980, pp. 341–64); and "The Role of the Vomeronasal Organ in Behavioral Control of Reproduction," in "Reproduction: A Behavioral and Neuroendocrine Perspective," *Annals of New York Academy of Sciences* 474: 148–57, 1986.
- Additional information on the mammalian vomeronasal organ can be found in J. G. Dulka, *Brain Behavioral Evolution* 42(4–5):265–80, 1993.
- Bombykol is discussed extensively in *Chemical Communication: The Language of Pheromones,* by William C. Agosta (Scientific American Library, 1992).
- The section on the mating behaviors of the male fruit fly refers to the article "Dance with Me, My Lovely," by L. Oliwenstein, in the January 1997 issue of *Discover.* It describes the work of evolutionary biologist William Rice of the University of California at Santa Cruz, who devised a way to create "supermale" fruit flies.
- For additional reading on tick pheromones, see *Biology of Ticks,* by D. E. Sonenshine, Vol. 1 (Oxford University Press, 1991, pp. 331–69); "Mounting Sex Pheromone: Its Role in Regulations of Mate Recognition in the Ixodidae," by D. E. Sonenshine, *Modern Acarology,* 1991, pp. 69–78; and "Chemically Mediated Behavior in the Acari: Adaptations for Finding Hosts and Mates," by D. E. Sonenshine, D. Taylor, and K. A. Carson, *Journal of Chemical Ecology* 12:1091–108, 1986.

- The description of how the Douglas fir beetle uses pheromones relies on information in the entry "Pheromone" presented in the Microsoft®Encarta® 1996 Encyclopedia, copyright © 1993–95, Microsoft Corporation, Funk & Wagnalls Corporation. The contributor is Carol Grant Gould.
- The section on cockroach mating behaviors is presented in the book *Life on a Little-Known Planet,* by Howard Ensign Evans (Lyons & Burford, 1993).
- The golden hamster's pheromone-directed mating behaviors are further described by William C. Agosta in *Chemical Communication: The Language of Pheromones* (Scientific American Library, 1992).
- Additional information on how vaginal secretions from female hamsters affect the males of the species can be found in *Physiological Behavior* 54(2):213–14, August 1993.
- See *Nature's Everyday Mysteries: A Field Guide to the World in Your Backyard,* by Sy Montgomery (Chapters Publishing Ltd., 1993), for details of the porcupine's breeding season.
- The information about male elephants in musth is from *Beastly Behaviors,* by Janine M. Benyus (Addison-Wesley, 1992).
- For more information on similarities between certain insect and elephant pheromones see "One Scent Woos Elephants and Insects," *Science News,* vol. 149, no. 10, p. 159, March 9, 1996.
- The description of the pheromone defense system of the sea anemone *Anthopleura elegantissima* is based on the work of Nathan Howe of Stanford University's Hopkins Marine Station. A summary of this topic appears in William C. Agosta's book *Chemical Communication: The Language of Pheromones* (Scientific American Library, 1992).
- The information on beavers comes from *A Natural History of Sex: The Ecology and Evolution of Mating Behavior,* by Adrian Forsyth (Chapters Publishing Ltd., 1986, 1993), and *Nature's Everyday Mysteries: A Field Guide to the World in Your Backyard,* by Sy Montgomery (Chapters Publishing Ltd., 1993).
- The description of ovulatory behavior among female marmosets is based on information from the article "A Monopoly on Maternity," by S. Richardson, which appeared in the February 1994 issue of *Discover.*
- Readers who want more information on spider pheromones can refer to the article "For Insects, the Buzz Is Chemical," by N. Angier, the *New York Times,* March 29, 1994. For a more scientific look into spider pheromones as they apply to *Linyphia triangularis,* a European variety of spider that lives primarily in forests, see "Identification of a Sex Pheromone from a Spider," by S. Schulz and S. Toft, *Science* vol. 260, June 11, 1993, pp. 1635–37.

- The passage on honeybees is based in part on "The Insect Mind: Physics or Metaphysics?" by J. L. Gould and C. G. Gould, published in the report *Animal Mind–Human Mind,* ed. D. R. Griffin (Berlin: Springer-Verlag, 1982). Additional information about honeybees can be found in *The Human Nature of Birds,* by Theodore Xenophon Barber, Ph.D. (St. Martin's Press, 1993); "Recognizing Relatives by Scent," *Science News,* vol. 149, no. 10, March 9, 1996; "How the Queen Bee Makes Her Pheromone," by Tina Adler, *Science News,* vol. 149, no. 13, March 30, 1996; and "Caste-Selective Pheromone Biosynthesis in Honeybees," by E. Plettner, K. Slessor, M. Winston, and J. Oliver, *Science* Vol. 271, No. 5257, p. 1851, March 29, 1996.
- The lives of ants are documented in great detail in *The Diversity of Life,* by E. O. Wilson (Harvard University Press, 1992).
- *When Elephants Weep: The Emotional Lives of Animals,* by Jeffrey Moussaieff Masson and Susan McCarthy (Delacorte Press, 1995).
- The sentence that begins "Roger Caras, the president . . ." refers to the book *The Intertwining Lives of Animals and Humans Throughout History,* by Roger A. Caras (Simon and Schuster, 1996).

Chapter 3. The Nose Knows: The Human Vomeronasal Organ

- The section that begins "The man whose work culminated . . ." refers to conversations with Dr. David Berliner conducted in Menlo Park, California, in January 1997, and by telephone on numerous occasions.
- Information about the presence of the human VNO relies on the following sources: Interviews with Dr. David Berliner, Dr. David Moran, and Dr. Luis Monti-Bloch; Human Pheromone Sciences, Inc., press release, March 1993; "The Functionality of the Human Vomeronasal Organ (VNO): Evidence for Steroid Receptors," by D. L. Berliner, L. Monti-Bloch, C. Jennings-White, and V. Diaz-Sanchez, *Journal of Steroid Biochemistry and Molecular Biology,* June 1996; "Sixth Sense Therapy Path to Be Reported," by Jerry Bishop, *Wall Street Journal,* April 11, 1996; and "Vomeronasal Epithelial Cells of the Adult Human Express Neuron-specific Molecules," by S. Takami, M. L. Getchell, Y. Chen, L. Monti-Bloch, D. L. Berliner, L. J. Stensaas, and T. V. Getchell, *NeuroReport* 4:375–78, April 1993.
- Additional information about the human vomeronasal organ and theories about the presence of the accessory olfactory bulb in adults can be found in "The Structure and Function of the Human Vomeronasal Organ," by D. T. Moran, L. J. Stensaas, L. Monti-Bloch, and D. L.

Berliner, in *Handbook of Olfaction and Gustation,* ed. R. L. Doty (Marcel Dekker, Inc., 1995, pp. 793–820).

- The passage that describes the frequency of the human VNO and its independence from the human sense of smell refers to the following sources: "The Vomeronasal (Jacobson's) Organ in Man: Ultrastructure and Frequency of Occurrence," by D. T. Moran, B. Jafek, and J. C. Rowley III, *Journal of Steroid Biochemistry and Molecular Biology* 39(4B):545–52, 1991; "Ultrastructure of the Vomeronasal Organ in Man: A Pilot Study," by D. T. Moran, B. Jafek, and J. C. Rowley III, presented at the Seventh Annual Meeting of the Association of Chemoreception Sciences, Sarasota, Florida, 1985; "Ultrastructure of the Human Vomeronasal Organ," by L. Stensaas, R. Lavker, L. Monti-Bloch, B. I. Grosser, and D. L. Berliner, *Journal of Steroid Biochemistry and Molecular Biology* 39(4B):553–60, 1991; and "The Sixth Sense," by Robert Taylor, *New Scientist,* January 25, 1997, pp. 36–40.
- The passages that describe experiments with vomeropherins and the human VNO rely on the following sources: "Effect of Putative Pheromones on the Electrical Activity of the Human Vomeronasal Organ and Olfactory Epithelium," by L. Monti-Bloch and B. I. Grosser, *Journal of Steroid Biochemistry and Molecular Biology* 39(4B):573–82, 1991; and "Nonolfactory Responses from the Nasal Cavity: Jacobson's Organ and the Trigeminal System," by D. Tucker, in "Olfaction," *Handbook of Sensory Physiology,* vol. 4, "Chemical Senses," part 1, ed. L. M. Beidler (Berlin: Springer-Verlag, 1971, pp. 151–81). For further reading about pheromones and their role in fragrances, see "Perfumery and the Sixth Sense," by C. Jennings-White, *Perfumer & Flavorist,* vol. 20, July/August 1995.
- The sentence that begins "As of this writing . . ." relies on a conversation with Dr. David Berliner.
- Additional reading on the human VNO can be found in the following articles that appeared in the popular press: "The Essence of Attraction," by Bruce Goldman, *Health,* March/April 1994; "A Sixth Sense That Affects How You Feel," by Gene Bylinsky, *Fortune,* January 27, 1992; "Pheromones: Messengers of Love," by Gerd Schuster, translated from the German, originally appearing in *Stern*, April 1996; and "The Sniff of Legend," by Karen Wright, *Discover,* April 1994.

Chapter 4. The Brain Dance of the Senses

- Information on the human brain in this chapter relies on the following sources: "Body Voyage: A 21st Century Anatomy Lesson," by Claudia

Glenn Dowling, *Life,* February 1997; *The Human Mind Explained: An Owner's Guide to the Mysteries of the Mind,* by Susan A. Greenfield, general editor (Henry Holt, 1996); "Finding Elusive Factors That Help Wire up Brain," by Natalie Angier, *New York Times,* August 16, 1994; "Mind Your Mind," by Peter Jaret, *Living Fit,* March 1996; "How the Brain Might Work: A New Theory of Consciousness," by Sandra Blakeslee, *New York Times,* March 21, 1995; "Mind and Brain: A Scientific Special Report," by Gerald D. Fischbach (Scientific American, Inc., 1992, 1994); and "Quiet Miracles of the Brain," by Joel L. Swerdlow, *National Geographic,* June 1995.

- Here are suggestions for additional reading on the human brain: *The Runaway Brain,* by Christopher Wills (Basic Books, 1993); *The Emotional Brain: The Mysterious Underpinnings of Emotional Life,* by Joseph LeDoux (Simon & Schuster, 1996); *Descartes' Error: Emotion, Reason and the Human Brain,* by Antonio R. Damasio (G. P. Putnam's Sons, 1994); and *Consciousness Explained,* by Daniel C. Dennett (Little Brown and Company, 1991).
- Some of the information regarding the anatomical structure of the brain is from *Gray's Anatomy,* by Henry Gray, F.R.S., drawings by H. V. Carter, M.D. (John W. Parker and Son, 1858; reprinted by The Promotional Reprint Company Limited, 1991); and "The Brain and the Cranial Nerve," in *Principles of Anatomy and Physiology,* by Gerald J. Tortora and Nicholas P. Anagnostakos (Harper & Row, 1987).
- Interesting reading on the historical aspects of the sense of touch can be found in the chapter "The Cutaneous Senses," in *Origins of Neuroscience,* by Stanley Finger (Oxford University Press, Inc., 1994).
- The paragraph that begins "The capacity of the eye . . ." refers to "Contrasting Views" in *The Human Mind Explained,* edited by Susan A. Greenfield (Henry Holt, 1996).
- The role of odor and smells in cultures around the world is based on The Olfactory Research Fund's fact sheet "The History and Anthropology of Smell."
- Sources for information on the human olfactory system include "The Fine Structure of the Olfactory Mucosa in Man," by David T. Moran, J. Carter Rowley III, Bruce W. Jafek, and Mark A. Lovell, *Journal of Neurocytology* 11:721–46, 1982; "Ultrastructural Studies of the Olfactory Neuroepithelium in Normal and Diseased States," by David T. Moran, presented to the fiftieth anniversary meeting of the American Academy of Allergy and Immunology, Chicago, March 13, 1993; "The Ultrastructure of the Human Olfactory Mucosa," by David T. Moran, Bruce W. Jafek, and J. Carter Rowley III, in the textbook *The Human Sense of Smell,* ed. D. Laing, R. L. Doty, and W. Breipohl (Springer-Verlag,

1992); "The Structure and Function of Sensory Cilia," by D. T. Moran and J. C. Rowley III, *Journal of Submicroscopic Cytology* 15:157–62, 1983; and "Electron Microscopy of Olfactory Epithelium in Two Patients with Anosmia," by D. T. Moran, B. W. Jafek, J. C. Rowley III, and P. M. Eller, *Archives of Otolaryngology,* 111:122–26, 1985.

- The passage that describes what happens when the olfactory epithelium is damaged is from conversations with Dr. David T. Moran; "The Loss of the Sense of Smell," an informational pamphlet published by The Olfactory Research Fund, 1996; and "What to Do When Your Patient Says: 'I Can't Smell Anything,' " by B. W. Jafek, B. A. Esses, and D. T. Moran, *Journal of Respiratory Diseases* 9:79–88, 1988.
- The sentence that begins "Fortunately, the neural tissues . . ." refers to "Consequences of Removing the Vomeronasal Organ," by Charles J. Wysocki and John J. Lepri, *Journal of Steroid Biochemistry and Molecular Biology* 39(4B):661–69, 1991; and "Cellular Interactions in Neuronal Development: Cell and Developmental Biology of Olfaction and Taste," abstract available on the Internet by Albert I. Farbman, Department of Neurobiology and Physiology, Northwestern University, October 1995.
- Additional information on olfaction can be found in the chapter "Olfaction and the Tracking Mouse," in *The Youngest Science,* by Lewis Thomas (Viking, 1983); and the chapter "Olfaction," in *Origins of Neuroscience,* by Stanley Finger (Oxford University Press, Inc., 1994).
- For more information on the five senses, see *The Five Senses,* by F. Gonzalez-Crussi (Harcourt Brace Jovanovich, 1989); *A Natural History of the Senses,* by Diane Ackerman (Random House, 1990); "Coming to Our Senses," by Shannon Brownlee with Traci Watson, *U.S. News & World Report,* January 13, 1997; "Taste Memory," by Molly O'Neill, *The New York Times Magazine,* May 12, 1996; *The Aroma-chology Review,* published by The Olfactory Research Fund, Inc., vol. IV, no. 1, 1995; and "Taste? Bud to Bud, Tongues May Differ," by John Willoughby, *New York Times,* December 7, 1994.

Chapter 5. Sex, Love, and Lust: The Pheromone Connection

- The passage that describes the sweaty T-shirt experiment refers to the following sources: *The Monell Connection* newsletter, Spring 1994 issue, published by the Monell Chemical Senses Center, a nonprofit scientific institute that conducts research on taste, smell, and chemosensory systems; and "Scent of a Man: The Sexiest Part of a Man, a Swiss Zoologist Has Found, May Be His Armpits," by Sarah Richardson, *Discover,* February 1996, pp. 26–28.

- The paragraph that begins "The Darwinian logic . . ." refers to a conversation with Dr. David T. Moran.
- The perspiration experiments conducted by Cutler and Preti are described in greater detail in *Love Cycles,* by Winnifred B. Cutler (Villard, 1991); and in "Human Axillary Secretions Influence Women's Menstrual Cycles: The Role of Donor Extract of Females," by G. Preti, W. B. Cutler, C. R. Garcia, G. R. Huggins, and H. J. Lawley, *Hormones and Behavior* 20:474–82.
- The source for information about the romance-inspired troubadours is from *The Power of Myth,* by Joseph Campbell with Bill Moyers (Doubleday, 1988, pp. 185–86).
- For more on why humans blush see "Why We Blush," a news brief in *Self* magazine, May 1996.
- The sources for information on phenylethylamine (PEA) are "Sweets, Chocolate, and Atypical Depressive Traits," by Schuman, Gitlin, and Fairbanks, *Journal of Nervous and Mental Disorders* 175:491–95, 1987; *The Chemistry of Love,* by Michael Liebowitz (Little Brown, 1983); *Anatomy of Love,* by Helen Fisher, Ph.D. (Ballantine, 1992); and *Chocolate to Morphine: Understanding Mind-active Drugs,* by A. Weil and W. Rosen (Houghton-Mifflin, 1983).
- The section on oxytocin is based on "Mating for Life? It's Not for the Birds or the Bees," by Natalie Angier, *New York Times,* August 21, 1990; and "A Potent Peptide Prompts an Urge to Cuddle," *New York Times,* C1–C10, 1991.
- Information on testosterone can be found in many sources, including: "Does Testosterone Equal Aggression? Maybe Not," by Natalie Angier, *New York Times,* June 20, 1995; "Male Hormone Molds Women, Too," by Natalie Angier, *New York Times News Service,* April 5, 1994; and "Male Hormone Molds Women, Too, in Mind and Body," by Natalie Angier, *New York Times,* May 3, 1994.
- The source for estrogen information is from the article "The Estrogen Clock," by John Sedgwick, *Self* magazine, December 1995, pp. 134–39.
- Dopamine and prosexual drugs are discussed in *The Alchemy of Love and Lust,* by Theresa Crenshaw, M.D. (Putnam, 1996).
- The sexual advantages of possessing symmetrical physical features are the topic of the following articles: "The Biology of Beauty," by Geoffrey Cowley, *Newsweek,* June 3, 1996, pp. 61–69; "Why Did You Pick Her? The Science of Love," by Bill Shapiro, *Men's Journal,* February 1997, p. 72; and "Beauty and the Beat," *Allure,* March 1996.
- The sentence that begins "Edwin Dobb writes . . ." refers to his article "A Kiss Is Still a Kiss," published in *Harper's,* February 1996, pp. 35–43.
- The passage that describes how kissing and other forms of affection can boost immune function is discussed in "Kiss & Well: How to

Smooch and Seduce Your Way to Health," by Jan Sheehan, *Longevity,* February 1996, p. 50.

- The section on kissing also relies on information from: *The One-Hour Orgasm: The Ultimate Guide to Totally Satisfying Any Man or Woman . . . Every Time,* by Bob and Leah Schwartz (Breakthru, 1995); *Sexational Secrets: Exotic Advice Your Mother Never Told You,* by Susan Crain Bakos (St. Martin's, 1996); *The Art of Kissing,* by William Cane (St. Martin's, 1995); and *The Great Divide: How Females and Males Really Differ,* by Daniel Evan Weiss (Simon & Schuster, 1991).

Chapter 6. More Pheromone Mysteries

- Numerous anecdotes about the sense of smell can be found in *Mammalian Olfaction, Reproductive Processes, and Behavior,* passage by Richard L. Doty, ed. R. L. Doty (Academic Press, 1976); and *The Sexual Life of Savages in North-western Melanesia,* by B. Malinowski (Routledge, 1929).
- Sources of information on apocrine glands include *The Apocrine Glands and the Breast,* by M. B. L. Craigmyle (John Wiley and Sons, 1984); and "The Intimate Sense of Smell," by B. Gibbons, *National Geographic* 170:324–61, 1986.
- Information on lovemaps refers to *Lovemaps: Clinical Concepts of Sexual/Erotic Health and Pathology, Paraphilia, and Gender Transposition in Childhood, Adolescence and Maturity,* by J. Money (Irvington Publishers, 1986).
- "The Ballad of Sexual Dependency" is featured, along with other essays on human sexuality, in *The Erotic Impulse: Honoring the Sensual Self,* ed. David Steinberg (Tarcher/Putnam, 1992).
- The section on oral sex is based on information from *A Natural History of Love,* by Diane Ackerman (Random House, 1994); *Sexuality and the Psychology of Love,* by Sigmund Freud, trans. Philip Rieff (Collier, 1963); and *Civilization and Its Discontents,* by Sigmund Freud, trans. James Strachey (W. W. Norton, 1961).
- Additional information on human body odors and pheromones is found in "The Hidden Power of Body Odors: Studies Find that Male Pheromones Are Good for Women," by John Leo, *Time,* December 1, 1986; and "The Chemistry Between People: Are Our Bodies Affected by Another Person's Scent?" by Terence Monmaney with Susan Katz, *Newsweek,* January 12, 1987.
- Wilhelm Fliess's studies of the menstrual cycle are described in *Scent: The Mysterious and Essential Powers of Smell,* by Annick Le Guérer (Turtle Bay Books, 1992).

- Information about copulins and their effects on rhesus monkeys is found in "Neuroendocrine Factors Regulating Primate Behaviour," by R. P. Michael, in *Frontiers in Neuroendocrinology,* ed. L. Martini and W. F. Ganong (Oxford University Press, 1971); "Chemical Signals and Primate Behavior," by R. P. Michael and R. W. Bonsall, in *Chemical Signals in Vertebrates,* ed. D. Muller-Schwarze and M. M. Mozell (Plenum Press, 1977).
- The sentence that describes the sexual-attraction methods of French prostitutes are described in *The Encyclopedia of Erotic Wisdom,* by R. C. Camphausen (Inner Traditions International, 1991).
- The passage on male facial hair refers to "Facts on Facial Hair," *New Woman,* September 1996, p. 38.
- Menstrual synchrony is described in "Menstrual Synchrony Between Mothers and Daughters and Between Roommates," by A. Weller and L. Weller, *Physiological Behavior* 53(5):943–49, 1993.
- Additional reading on menstrual synchrony can be found in "Menstrual Synchrony in Female Undergraduates Living on a Coeducational Campus," by C. A. Graham and W. C. McGrew, *Psychoneuroendocrinology* 5:245–52, 1980; "Menstrual Synchrony and Suppression," by M. McClintock, *Nature* 229:244–45, 1971; and "Olfactory Influences on the Human Menstrual Cycle," by M. J. Russell, G. M. Switz, and K. Thompson, *Pharmacology Biochemistry and Behavior* 13:737–38, 1980.
- The passage that discusses studies involving human pheromones and how they can affect a person's social interactions are discussed in "Irritability and Depression During the Menstrual Cycle—Possible Role for Exogenous Pheromone?" by J. J. Cowley, F. Harvey, A. L. Johnson, and B. W. L. Brooksbank, *Irish Journal of Psychology* 4:143–56, 1980; "Human Exposure to Putative Pheromones and Changes in Aspects of Social Behaviour," by J. J. Cowley and B. W. L. Brooksbank, *Journal of Steroid Biochemistry and Molecular Biology* 39(4B):647–59, 1991; "The Likelihood of Human Pheromones," by A. Comfort, *Nature* 230:432–33, 1971; and "The Influence of Androstenol—a Putative Human Pheromone—on Mood Throughout the Menstrual Cycle," by D. Benton, *Biological Psychology* 25:249–56, 1982.
- The passage that describes how pheromones might affect a person's opinion of and feelings toward other people (in this case, political candidates) is described in "The Effect of Two Odorous Compounds on Performance in an Assessment-of-People Test," by J. J. Cowley, A. L. Johnson, and B. W. L. Brooksbank, *Psychoneuroendocrinology* 2:159–72, 1977.
- The section that discusses personality differences between people living in warm and cold climates are discussed in "It's True: Hot Climate Makes Warm People," *Health,* July/August 1996.

- The sentence that begins "Charles Darwin noted that an infant . . ." refers to information provided in *Chemical Communication: The Language of Pheromones,* by William C. Agosta (Scientific American Library, 1992).
- The passage that discusses mother-baby bonding is based on "Recognition of Maternal Axillary Odors by Infants," by J. M. Cernoch and R. H. Porter, *Child Development* 56:1593–98, 1985; "Olfaction in the Development of Social Preferences in the Human Neonate," by A. MacFarlane, *Ciba Foundation Symposium* 33:103–13, 1975; and "Human Olfactory Communication," by M. J. Russell, *Nature* 260:520–22, 1976.
- The paragraph that describes the pheromonal bond between mothers and babies in the animal world is described in "Sexual Attractivity, Proceptivity, and Receptivity in Female Mammals," by F. A. Beach, *Hormones and Behavior* 7:105–38, 1976.
- The passage that begins "However, pheromones can have negative effects . . ." refers to, "Pheromonal Emission by Pregnant Rats Protects Against Infanticide by Nulliparous Conspecifics," by J. A. Mennella and H. Moltz, *Physiology and Behavior* 46:591–95, 1989.
- The passage that describes a possible pheromonal connection to homosexuality is based on information from *Queer Science: The Use and Abuse of Research into Homosexuality,* by S. LeVay (MIT Press, 1996); *A Separate Creation: The Search for the Biological Origins of Sexual Orientation,* by C. Burr (Hyperion, 1996); *Homosexuality: A Philosophical Inquiry,* by M. Ruse (Oxford, 1988); "Homosexuality: A New Endocrine Correlate," by M. Margolese, *Hormones and Behavior* 1:151, 1970; and "Androsterone-etiocholanolone Ratios in Male Homosexuals," by M. Margolese and O. Janiger, *British Medical Journal* 207:207–10, 1973.
- The passage that describes violence prediction is based on information in the article "The Smart New Way to Protect Yourself from Crime," by Janis Graham, *Redbook,* July 1997.
- The discussion of Kallmann's syndrome is based on information from "Delayed Puberty, Eroticism and Sense of Smell: A Psychological Study of Hypogonadotropinism, Osmatic and Anosmatic (Kallmann's Syndrome)," N. A. Bobrow, J. Money, and V. J. Lewis, *Archives of Sexual Behavior* 1:329–44, 1971; "Kallmann's Syndrome: From Genetics to Neurobiology," by E. I. Rugarli and A. B. Ballabio, *Journal of the American Medical Association* 270:2713–16, 1993; and the abstract "Absence of Vomeronasal Organ (VNO) Function in Patients with Hypogonadotropic Hypogonadism," by L. Monti-Bloch, V. Diaz-Sanchez, and D. L. Berliner, Dept. of Psychiatry, University of Utah, the Department of Reproductive Biology, Instituto Nacional de la Nutricion, Mexico City, Mexico, and Pherin Pharmaceuticals, Menlo Park, California.

- The discussion of what happens when the vomeronasal organ of a rodent is removed is based on: "Release of LH in the Female Rat by Olfactory Stimuli: Effect of the Removal of the Vomeronasal Organs or Lesioning of the Accessory Olfactory Bulbs," by C. Beltramino and S. Talesnik, *Neuroendocrinology* 36:53–58, 1983; and "Neurobehavioral Evidence for the Involvement of the Vomeronasal System in Mammalian Reproduction," by C. J. Wysocki, *Neurosci. Biobehav. Rev.* 3:301–41, 1979.
- The paragraph that begins "Pheromone researchers Charles Wysocki . . ." refers to the research paper, "Consequences of Removing the Vomeronasal Organ," by C. J. Wysocki and J. Lepri, *Journal of Steroid Biochemistry and Molecular Biology* 39(4B):661–69, 1991. How the pheromones and body odors of male rodents affect the fertility of the female of the species is recounted in "Menstrual Synchrony and Suppression," by Martha McClintock, *Nature* 229:244–45, 1971; and "Pheromonal Regulation of the Ovarian Cycle: Enhancement, Suppression, and Synchrony," by Martha McClintock, in *Pheromones and Reproduction in Mammals,* ed. J. Vandenbergh (Academic Press, 1983, pp. 113–49).
- The section that describes the importance of preserving the VNO during nasal surgery is detailed in "The Incidence of the Vomeronasal Organ in 1000 Human Subjects and Its Possible Clinical Significance," by J. Garcia-Velasco and M. Mondragon, *Journal of Steroid Biochemistry and Molecular Biology* 39(4B):561–63, 1991; and "Nose Surgery and the Vomeronasal Organ," by J. Garcia-Velasco and S. Garcia-Casas, *Aesthetic Plastic Surgery* 19:451–54, 1995.

Chapter 7. Love Chemistry 101: Perfume and the Sixth Sense

- Much of the information for this chapter comes from conversations with Dr. David Moran, Dr. David Berliner, and other pheromone researchers, as well as from informational material obtained directly from Human Pheromone Sciences, Inc., the producer of the Realm pheromone fragrance described in this chapter. For more information on pheromones in colognes and perfumes read "The Human Skin: Fragrances and Pheromones," by D. L. Berliner, C. Jennings-White, and R. M. Lavker, *Journal of Steroid Biochemistry and Molecular Biology* 39(4B):671–79, 1991.
- The Sense of Smell Survey we refer to was a joint effort of the *New York Times* Marketing Research Department and the Olfactory Research Fund. It took place on June 5–18, 1995, and involved telephone interviews with 1,002 men and women. The results were tabulated by In-

tersearch Corp. and featured in a special advertising section in the October 22, 1995, issue of the *New York Times Magazine.*

- The paragraph describing the fragrances that can supposedly cure an ailing love life are described in *Love Magic,* by Marina Medici (Simon & Schuster, 1994).
- Numerous articles and books focus on the vast topic of perfume and fragrance. For the history of fragrance, we relied on the following sources: "Aromatherapy," *First,* December 18, 1995; *The Foul and the Fragrant,* by Alain Corbin (Harvard University Press, 1986); *A Natural History of the Senses,* by Diane Ackerman (Random House, 1990); *Fragrance,* by Edwin T. Morris (Scribner's, 1986); and the novel *Perfume,* by Patrick Süskind (Washington Square Press, 1991).
- The paragraph on the ancient Greeks relies on information from the book *Down-to-Earth Beauty,* by Catherine Palmer (St. Martin's Press, 1981).
- The sentence that begins "Smells enhance memories . . ." refers to information in the article "Design Credo: Heed the Nose," by Mitchell Owens, *New York Times,* June 16, 1994.
- The passage that describes how Japan's PuroLand theme park and the Mirage Hotel in Las Vegas use environmental fragrance is based on the article "Scents That Alter Your Moods," by Rona Berg, *Self,* December 1995, p. 146.
- For a more detailed discussion of how odors can affect learning and recall, see "The Retro-active Effect of Pleasant and Unpleasant Odors on Learning," by J. B. Frank and E. J. Ludvigh, *American Journal of Psychology,* vol. 43, pp. 102–8; "Effects of Ambient Odours of Lavender and Cloves on Cognition, Memory, Affect and Mood," by H. W. Ludvigson and T. R. Rottmann, *Chemical Senses,* vol. 14, pp. 525–36; and "The Effect of Smell on Cognitive Processes," by K. H. Berg, *Dragoco Report,* vol. 39, pp. 128–29.
- How vanilla fragrance can reduce stress and anxiety is discussed in the article "Nosing Around," by Janice Min, *Allure,* April 1997, p. 130.
- The discussion of the fragrance studies of Dr. Alan Hirsch is based on a taped interview conducted in January 1994 by Laura Lee of the *Laura Lee Radio Show* and reprinted in the June 1995 issue of the *Townsend Letter for Doctors.* For more information about Hirsch's work, see "Aroma-Chology: A Status Review," by J. Stephan Jellinek, reprinted in *Perfumer & Flavorist,* vol. 19, September/October 1994.
- How scents can affect shoppers and museum-goers is discussed in "Ambient Odour and Shopping Behaviour," by S. C. Knasko, *Chemical Senses,* vol. 14; and "Lingering Time in a Museum in the Presence of Congruent and Incongruent Odours," by S. C. Knasko, *Chemical Senses,* vol. 18.

- The information about Chinese courtesans and musk-flavored foods is found in the article "Scents of Desire," by Charlotte-Anne Fidler, British *Elle,* December 1996, p. 172.
- The paragraph that begins "Even today, as one fragrance study revealed . . ." refers to information in the *H&R Fragrance Guide,* Johnson, 1985.
- The sentence that begins "In France, one perfume . . ." refers to information provided in the book, *Scent,* by Annick Le Guérer (Turtle Bay Books, 1992).
- Clive Jennings-White's pheromone and perfume studies are summarized in "Perfumery and the Sixth Sense," *Perfumer & Flavorist,* vol. 20, July/August 1995.
- The information and direct advertising quotes for the products described in the section "Snake Oil?" were obtained directly from the Internet. Information on Athena Pheromone 10:13 was provided by the Athena Institute for Women's Wellness and via a telephone conversation with Athena's Winnifred Cutler, Ph.D.

Chapter 8. The Future of Pheromones

- Much of the information in this chapter was gathered during three years of interviews with scientists engaged in pheromone research. Since we received the information directly from these sources, we cannot provide citations for some of the information contained in this chapter. However, as we found it useful to access articles in the popular press that pertain to certain topics in this chapter, we do provide those references. Also, it is important to note that additional information was obtained from Pherin Pharmaceuticals, Inc., reports that are not available to the public. We were granted access to these documents and other highly confidential scientific information on the condition that we not reveal details that could potentially jeopardize Pherin's patent application process and its pharmaceutical research and development.
- The paragraph that beings "Pharmaceuticals are big business . . ." refers to the article "A Giant Battles Its Drug Dependency," by Richard Evans, *Fortune,* August 5, 1996.
- The passage that begins "Pherin Pharmaceuticals, Inc., was founded . . ." refers to conversations with Dr. David Berliner and information from internal Pherin reports.
- The passage that begins "Enter Pherin and its team of scientists . . ." refers to a Pherin press release dated October 19, 1995.
- For more details on how vomeropherins can affect the human vomeronasal organ consult "The Functionality of the Human Vomeronasal Organ (VNO): Evidence for Steroid Receptors," by D. L.

Berliner, L. Monti-Bloch, C. Jennings-White, and V. Diaz-Sanchez, *Journal of Steroid Biochemistry and Molecular Biology,* June 1996.

- Information about luteinizing hormone and its effects on the hypothalamus-pituitary link is based on Pherin reports and overviews of vomeropherin technology, as well as conversations with Dr. David Moran.
- The description of the world wide web's Testosterone Source is found in "Can Men Have too Little Testosterone?" *New Age,* July 1997.
- The sentence that begins "Prostate cancer is the second leading cause . . ." refers to information provided in the article "Twelve Major Cancers," *Scientific American,* September 1996. More information about prostate cancer can be found in "They're Closing in on Prostate Cancer," by Sally Squires, *Reader's Digest,* January 1997; "Evidence Links Flawed Gene to Some Prostate Cancers," by Paul Recer, The Associated Press, November 22, 1996; "Scientists Zero in on Gene Tied to Prostate Cancer," by Natalie Angier, *New York Times,* November 22, 1996; "Attention: Aging Men," by Geoffrey Cowley, *Newsweek,* September 16, 1996; "Clues to Prostate Cancer," by Bonnie Liebman, *Nutrition Action Newsletter,* March 1996; "Does Screening for Prostate Cancer Make Sense?" by Gerald E. Hanks and Peter T. Scardino, *Scientific American* magazine, September 1996; "The most deadly cancers," Special Report, *U.S. News & World Report,* February 5, 1996; "To Screen, or Not to Screen," *The Economist,* March 29, 1997; and "Europe Prostate Drug Rights Are Sold to Unit of L'Oréal," by Lawrence M. Fisher, *New York Times,* June 2, 1997.
- Information about prostate cancer treatments is provided in "Early Prostate Surgery Is Found Very Effective," *New York Times,* August 28, 1996.
- Facts about anxiety and panic attacks are from the American Academy of Family Physicians' patient information brochure titled "Anxiety and Panic: Gaining Control Over How You're Feeling," February 1996. For more information about panic disorder, read "Night of the Living Dread," by Mark Kram, *Men's Health,* April 1997. Sources for further reading on anxiety include the following: "Happier Days Ahead? The Future of Mind Drugs," by Carla Koehl, *Newsweek,* April 21, 1997; "Anxiety Linked to Gene Affecting Brain Chemistry," by Paul Recer, The Associated Press, November 29, 1996; "Grumpy, Fearful Neurotics Appear to Be Short on a Gene," by Natalie Angier, *New York Times,* November 29, 1996; "Don't Worry, Be Healthy?" by Jan Jarboe Russell, *Mirabella,* January 1997; "How to Beat Anxiety," *First,* January 20, 1997; "Oh, How Happy We Will Be: Pills, Paradise, and the Profits of the Drug Companies," by Greg Critser, *Harper's,* June 1996.

- Our obsession with thinness is discussed in "Gaining on Fat," by W. Wayt Gibbs, *Scientific American,* August 1996. For more information on the diet drug industry and the business of weight loss, consult "Dieting Dangerously," by Richard Klein, *New York Times,* July 14, 1997; "The Chemical of Craving," by Hara Estroff Marano, *Self,* July 1997; "The Fearful Price of Getting Thin," by Gina Kolata, *New York Times,* July 13, 1997; "Diet-drug Combo Can Pose Risks," The Associated Press, July 9, 1997; "Can a Pill Make You Thin?" by Laurie Tarkan, *Family Circle,* August 6, 1996; "Here We Go Again: The Newest Diet Drug Is Only for the Obese, So Why Are the Rest of Us So Interested?" *Elle,* August 1996; "Hunger for Less: Next-wave Diet Pills," by Jim Thornton, *Men's Journal,* December 1996/January 1997; "What Price Diet?" *Mademoiselle,* August 1995; "Obesity Epidemic," by Susan Learner Barr, *Shape,* October 1996; "Killer Diets: The New Fat-pill Junkies," by Douglas S. Barasch, *Cosmopolitan,* April 1997.
- Additional sources for information on Redux and other diet drugs are "Diet Redux: The Latest Diet Drug Was Designed to Fight Extreme Obesity, Not Help You Fit into Tight Jeans," by Andrea Barnet, *Self,* October 1996; "The Pill Mill," by Jeannie Ralston, *Allure,* July 1997; "The New Miracle Drug?" by Michael D. Lemonick, *Time,* September 23, 1996; "Aided by Newer Products, 3 Drug Makers Post Higher Profits," The Associated Press, as published in the *New York Times,* July 24, 1996; and "Two Popular Diet Pills Linked to Problems with Heart Valves," by Gina Kolata, *New York Times,* July 9, 1997.
- The passage describing the pheromone pest-control work of Cornell University chemist Dr. Jerrold Meinwald relies on information provided in the July 1995 issue of *Science* magazine.
- Additional information on pheromone pest-control formulas can be found in the May 1995 issue of *Fruit Grower* in the article "Codling Moths Do It . . . Twig Borers Do It . . . Now Filbert Nut Moths Might Start Romancing Pheromones," by K. Morley-Smith; see also "Novel Cockroach Chemistry," by Julie Ann Miller, *BioScience,* vol. 43, no. 5, May 1993.
- The sentence that begins "One Pherin investor . . ." refers to a Pherin letter to shareholders dated August 12, 1996.
- Pherin's agreement with Akzo Nobel NV/NV Organon is described in "Birds Do It, Bees Do It . . . Now Akzo Bets We Do It: Steer by Pheromones," by Ron Winslow, *Wall Street Journal,* July 21, 1997; and to multiple conversations with Dr. David Berliner.

GLOSSARY

Alpha waves: Brain waves (recorded in an electroencephalogram, or EEG) that are regular, characteristic of the relaxed state.

Amygdala: A small area in the temporal lobe of the brain that has connections with the olfactory system, the limbic system, and the hypothalamus.

Anatomy: The study of the structure of a specific kind of living organism.

Anosmia: Total lack of the sense of smell.

Anxiety: An uncomfortable state of emotional and physiologic tension, often closely associated with fear, that can occur in response to a physical, psychological, or pharmacologic stimulus.

Anxiolytic: A substance that reduces anxiety and the feelings associated with it.

Aphrodisiac: A substance—named for Aphrodite, the Greek goddess of love—that stimulates intense sexual desire in one who partakes of it.

Apocrine glands: Large secretory glands located in the skin and concentrated near the nipples, armpits, and groin that secrete a pheromone-rich viscous substance in times of stress or excitement. Apocrine glands produce the heavy "sweat" associated with body odor.

Aromatherapy: The practice of bringing about a change in feelings by exposure to specific odors. Aromatherapy, which works through the sense of smell, is not intended to involve pheromones and the sixth sense.

Autonomic: Bodily functions believed to be independent of conscious control. Example: The constriction of surface blood vessels that occurs in response to cold is an autonomic function related to regulation of body temperature.

Benzodiazepines: Any of a group of tranquilizers, the best known of which is Valium (diazepam), which have antianxiety, muscle-relaxing, and sedative effects.

Beta waves: Brain waves (recorded in an electroencephalogram) that occur during periods of intense activity of the nervous system.

Biochemistry: The study of the chemical reactions that occur in living things.

Bioelectrical impulses: Electrical impulses carried by specialized cells such as nerve and muscle fibers.

Biotechnology: The use of modern technology in concert with existing biological materials or processes to achieve a unique product or result.

Bombykol: The pheromone produced by the female silkworm moth, *Bombyx mori*, which attracts male moths of the same species (but not other kinds of moths) from afar for mating.

Cell: A unit of biological structure, of which all living things (save viruses and bacteria) are made, consisting of a nucleus, organelles, and cytoplasm, all surrounded by a thin, highly functional cell membrane.

Cineole: A compound obtained from eucalyptus oil with a strong, camphor-like odor.

Civet: A strong-smelling substance derived from the anal glands of the civet cat frequently used in making perfume.

Compound: A pure substance composed of a single kind of molecule that, in turn, is made of more than one kind of atom. Example: Sodium and chlorine are elements; when combined, they form the compound sodium chloride, or salt.

Cortex (cerebral): The thin layer of gray matter on the surface of the brain associated with higher mental functions in man.

Depression: A psychological state often caused by repressed emotions, such as anger, in which the depressed person feels sad, tired, and has difficulty performing ordinary daily tasks.

Dopamine: A neurotransmitter in the central nervous system.

Eccrine glands: Small glands located in the skin that produce the watery sweat that lowers the body temperature by evaporative cooling.

EEG: An electroencephalogram obtained by monitoring overall brain activity by means of electrodes attached to the surface of the scalp.

Electron microscope: A powerful microscope using an electron beam to produce an image that can detect objects far too small to be seen with a conventional light microscope.

Emotion: A strong feeling experienced by a person or other animal, often accompanied by (or in response to) a physiologic change.

Endocrine system: A system of "ductless" glands, such as the thyroid, pituitary, and adrenals, that secrete hormones that enter the bloodstream and are transported to target organs, upon whose function they have a dramatic effect.

Endocrinology: The branch of biomedical science centered on the study of the endocrine system.

Epithelium: A sheet of cells, one or more cell layers thick, that covers a tissue or an organ. The uppermost part of the human skin—the epidermis—is an epithelium, as is the surface of the lining of the intestine.

Eros: The god of love in Greek mythology—known as Cupid in Roman mythology—often depicted as an immature winged angel armed with a bow and arrow, whose arrows inflict their human targets with powerful feelings of love and passion.

Estrogen: A female hormone produced by the ovaries that promotes the formation and maintenance of secondary sexual characteristics in women.

Estrus: The time in which a female mammal is sexually ready. Also called the heat or rut.

Food and Drug Administration (FDA): An arm of the U.S. government that sets rigorous standards for safety in foods and drugs made available for sale on the U.S. market.

Galvanic skin response: The measurement of the skin's ability to conduct an electrical current across its surface. A parameter used in lie-detector tests.

Histochemistry: The study of the composition of cells and tissues, determined by microscopic examination of histologic preparations (slides) stained with specific (histochemical) agents.

Homeostasis: The process by which one's metabolic activities are kept "on an even keel," leading to a state of health.

Hormone: A physiologic organic substance secreted by living cells that enters the bloodstream and travels to a target organ or organs wherein it brings about a physiologic adjustment in accord with the body's needs.

Hypogonadism: A condition in which the gonads are small and incompletely developed.

Hypothalamus: Often called "the brain's brain," the hypothalamus is that part of the brain that controls basic functions such as endocrine activity, water balance, body temperature, sleep, appetite, and sex drive.

Impotence: The inability to achieve an erection.

Intuition: An instinctive mental process by which you "automatically" realize a truth—often about another person—without resorting to rational thought.

Invertebrate: An animal, such as a crab, mosquito, or clam, that has no backbone.

Jacobson's organ: Another name for the vomeronasal organ. Jacobson was a scientist who gave one of the early descriptions of the vomeronasal organ in nonhuman mammals.

Kallmann's syndrome: A genetic condition characterized by incomplete sexual development, sterility, lack of sense of smell, and lack of a functioning vomeronasal organ. The olfactory bulbs of the brain are missing in people with Kallmann's syndrome.

Libido: Sexual appetite.

Limbic system: A part of the brain associated with, among other things, olfaction and emotions.

Lordosis: A position of the body in which the rear end of the animal is raised, often adopted by female vertebrates during copulation.

Luteinizing hormone (LH): A hormone made by the pituitary gland that promotes development of the corpus luteum in a woman's ovaries during pregnancy. The corpus luteum, in turn, secretes progesterone.

Luteinizing hormone-releasing hormone (LHRH): A releasing factor made and secreted by the brain's hypothalamus that causes the pituitary gland to release luteinizing hormone.

Menstrual synchrony: Occurs when the menstrual cycles in a group of women become synchronous (happen at the same time).

Metabolism: The sum total of ongoing biochemical reactions that maintain life in an organism.

Microgram: A unit of weight equal to one millionth of a gram (.000001 gram).

Micrograph: A photograph taken with a microscope.

Molecule: A structure formed by a combination of atoms. Molecules can range from simple ones (such as water) to complex ones (such as cholesterol and pheromones).

Nanogram: A unit of weight equal to one billionth of a gram (.000000001 gram).

Nasal septum: The thin sheet of bone and cartilage, covered by a mucous membrane, that runs down the midline of the nasal cavity, dividing it into two bilaterally symmetrical halves.

Nervous system: The incredibly complex communication system in the body made up of nerve cells and all their interconnections.

Neural pathway: A specific route, taken by a nerve or group of nerves, that conducts electrical excitation from one part of the body to another. Example: The olfactory nerve is the neural pathway by which olfactory receptor neurons, stimulated by odors, carry their electrical excitation to the olfactory bulbs of the brain.

Neuroanatomy: The study of the human brain and spinal cord.

Neuron: A nerve cell. Neurons take many shapes and have many functions.

Odor: Any scent or smell detected by the olfactory receptors in the nose.

Olfactory bulbs: A pair of small projections at the base of the brain that receive input from the olfactory nerves (which, in turn, contain nerve fibers originating in olfactory sensory neurons in the nasal cavity).

Olfactory mucosa: The sensory end organ of the olfactory system. Located in the upper reaches of the nasal cavity, the human olfactory mucosa contain the sensory receptor neurons that, when stimulated by odorant molecules, send signals to the olfactory bulbs in the brain via the olfactory nerve.

Olfactory system: The sensory system that mediates the sense of smell.

Organic: A substance derived from living matter that is usually a carbon compound.

Organic chemistry: The study of carbon compounds. Organic molecules, whose structure is based on chains or rings of carbon atoms, often contain hydrogen, oxygen, and nitrogen.

Patent: A legal device by which one's proprietary interest in an invention or discovery can be protected for a certain period of time.

Pharmacology: The study of the ways in which specific drugs affect humans and other organisms.

Pheromone: A chemical messenger released by an individual and perceived by another individual (or individuals) of the same species.

Physiology: The study of the body's chemistry and the ways in which specific cells and tissues perform the functions of life.

Pituitary gland: A pea-sized gland (sometimes called the master gland) located at the base of the brain that secretes hormones into the bloodstream, which affect such important functions as growth, metabolism, sexual maturation, the menstrual cycle, activity of the adrenal gland, and salt and water balance. The release of many hormones from the pituitary gland is controlled by the hypothalamus.

Premenstrual syndrome: A time of distress and discomfort, manifest in varying degrees in different women, that may occur during the ten days preceding menstruation.

Progesterone: A female hormone produced by the ovaries during pregnancy that prevents further ovulation.

Prostate: A gland located near the bladder in men that secretes a viscous fluid into the urethra at the time of ejaculation. Prostatic fluid forms the bulk of the ejaculate.

Psychophysiology: The science that deals with the relationship between psychologic and physiologic processes.

Psychotropic: Exerting an effect upon the mind; usually, a drug that affects one's mental state.

Puberty: The onset of sexual maturation, usually marking the beginning of adolescence.

Receptor cells: Specialized nerve cells—sensory neurons—adapted to respond to an environmental stimulus. Examples: Photoreceptor cells in the eye's retina are stimulated by light; chemoreceptor cells in the VNO are stimulated by pheromones.

Sebaceous glands: Large secretory glands located in the skin and usually associated with hair follicles that produce a waxy, yellowish secretion that lubricates hairs and certain regions of the skin such as the nose and the lining of the ear canals.

Sensory: Something influenced by the senses, in which an environmental stimulus activates a sensory receptor, bringing about an electrical response in associated nerves.

Septal deviation: A defect in the nasal septum in which that structure, pushed out of place and to one side, interferes with normal nose breathing.

Serotonin: A molecule made by the body that can cause blood vessels to constrict. It also acts as a neurotransmitter in the brain.

Species-specific: An adjective applied to a substance or behavior that, when produced or displayed by a member of one species, has an effect on another member of the same species. Pheromones are species-specific: Human pheromones affect other humans, but not pigs; pig pheromones affect other pigs, but not humans.

Sympathetic nervous system: The part of the involuntary (autonomic) nervous system that promotes the "fight or flight" response to threatening situations.

Synthesis: The process by which a molecule is assembled.

Systemic absorption: The absorption of materials, such as nutrients and drugs, into the body by way of the digestive tract.

Testes: The testicles, the male organs of sperm and androgen production.

Testosterone: A male hormone, or androgen, produced in the testes that stimulates the formation and maintenance of secondary sexual characteristics in men. Testosterone is a steroid hormone—one based on the cholesterol molecule.

Thalamus: Part of the brain that relays sensory information to the cerebral cortex.

Toxicity: A toxic substance is poisonous; toxicity refers to how poisonous it is.

Transdermal patch: An adhesive skin patch containing a drug that slowly delivers that drug across the skin and into the circulation.

Vertebrate: An animal, such as a human, cat, or giraffe, that has a backbone.

Vestigial: A structure, usually an organ, that is present in reduced form but has no known function essential for the life of the animal. The human appendix is usually considered to be a vestigial organ.

Volatile: A substance that evaporates readily, often giving off molecules that stimulate the sense of smell or the sixth sense (the pheromone sense).

Vomeronasal organ (VNO): The sense organ of the sixth sense that detects pheromones. In humans, it is a pair of organs, one on each side of the lower aspect of the nasal septum near the rear of the nostril, that responds to pheromones and sends sensory information to the brain.

Vomeropherin: A chemical substance that stimulates the human vomeronasal organ. It may be a naturally occurring pheromone and/or a compound synthesized in the laboratory.

INDEX